Solar System
PHYSC 159

Professor Laurel Senft

Published by SUNY OER Services
Milne Library
State University of New York at Geneseo
Geneseo, NY 14454

Distributed by State University of New York Press

Cover design by Elaine Farrally-Plourde.
Cover images in order from left to right provided by skeeze at https://pixabay.com/en/solar-system-montage-planets-space-639582/, Imagine_Images at https://pixabay.com/en/planet-astronomy-moon-science-3149121/, Bryan Minear at https://unsplash.com/photos/UglVBZGpaFg, Engin_Akyurt at https://pixabay.com/en/abstract-light-color-color-image-2408490/, John Reed at https://unsplash.com/photos/qtRFE7MYnHo, WikiImages at https://pixabay.com/en/neptune-planet-solar-system-space-11629/, Comfreak at https://pixabay.com/en/solar-system-sun-mercury-venus-439046/, and GuillaumePreat at https://pixabay.com/en/space-planet-earth-orbit-sky-1201086/.

ISBN: 978-1-64176-045-4

Table of Contents

Chapter 1: Classifying Solar System Objects

In this chapter, you will learn how scientists classify objects in the solar system. You will see how scientists define a planet, what the difference between a Terrestrial planet and a Jovian planet is, and how other objects, like dwarf planets, are defined.

As you read this, you are sitting somewhere on the planet Earth. Earth is one of the eight planets that orbit the Sun in our solar system. In order from closest to the Sun to furthest, those planets are Mercury, Venus, Earth, Mars, Jupiter, Saturn, Uranus, and Neptune. At one time Pluto, which orbits beyond Neptune, was also considered a planet; however it is no longer considered so. We will see why shortly. Our Sun is but one of hundreds of billions of stars which make up our galaxy, the Milky Way Galaxy. Finally, our galaxy is but one of hundreds of billions of galaxies in the universe!

Of course, ancient Greek and Roman astronomers did not know this. All these astronomers knew was that when they looked up at the sky, there were pinpricks of light. They noted that these pinpricks of light remained in the same patterns and did not move relative to one another (this is why we can define constellations – because the stars that make up the Big Dipper, for example, are always in the same place in reference to one another). However, they also noted that there were a few pinpricks of light that did not obey this rule; these pinpricks of light moved around against the background pattern of stars. Thus they called these pinpricks of light "planets" (meaning "wandering star" in Greek) and named them Mercury, Venus, Mars, Jupiter, and Saturn after their gods.

As science progressed, scientists realized that planets are actually balls of rock and/or gas that orbit the Sun. And as telescopes were invented and advanced, we discovered three more planets: Uranus in 1781, Neptune in 1846, and Pluto in 1930.

The problem (for Pluto!) arose around the turn of the millennium, when scientists suddenly began to find many more objects similar in size to Pluto (some even bigger) and in similar orbits as Pluto.

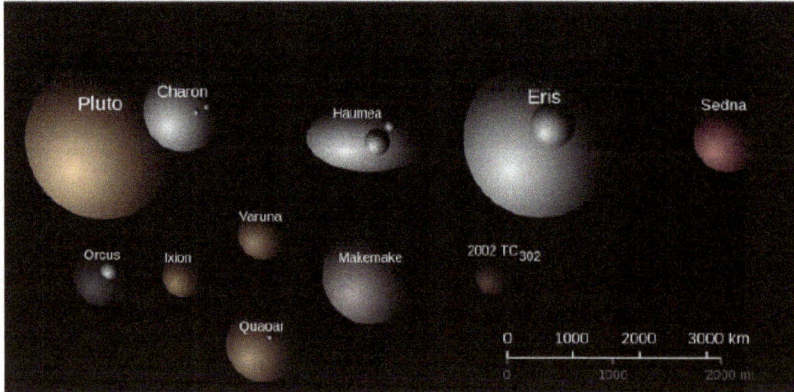

Some of these objects are shown in Figure 1.1. Scientists were faced with a decision: either they needed to call all of these new objects planets as well, or they had to change the definition of a planet. They choose the later, and in 2006 the International Astronomical Union presented the new scientifically accepted definition of a planet.

Figure 1.1: Solar system objects with similar sizes and orbits as Pluto. (https://commons.wikimedia.org/wiki/File:TheTransneptunians_Size _Albedo_Color.svg)

According to this new definition, a planet is an object that (1) orbits the Sun, (2) is big enough to be round (> 1,000 km in diameter), and (3) has cleared its orbit of debris (think of a truck driving down the highway alone, versus a truck surrounded by other trucks and cars!). Pluto orbits the

Sun and is big enough to be round, but because many other objects surround it, it has not cleared its orbit of debris. Thus Pluto is no longer considered a planet.

There are two types of planets: Terrestrial and Jovian. The differences can be seen on the plot in Figure 1.2 that shows the densities, sizes, and compositions of the planets. The Terrestrial planets are Mercury, Venus, Earth, and Mars. These are relatively small, rocky, high density planets with orbits that are closer to the Sun. The Jovian planets are Jupiter, Saturn, Uranus, and Neptune. These are relatively large, gaseous, low density planets with orbits that are further from the Sun. Unlike the Terrestrial planets, the Jovian planets have no solid surface, but they do have rings.

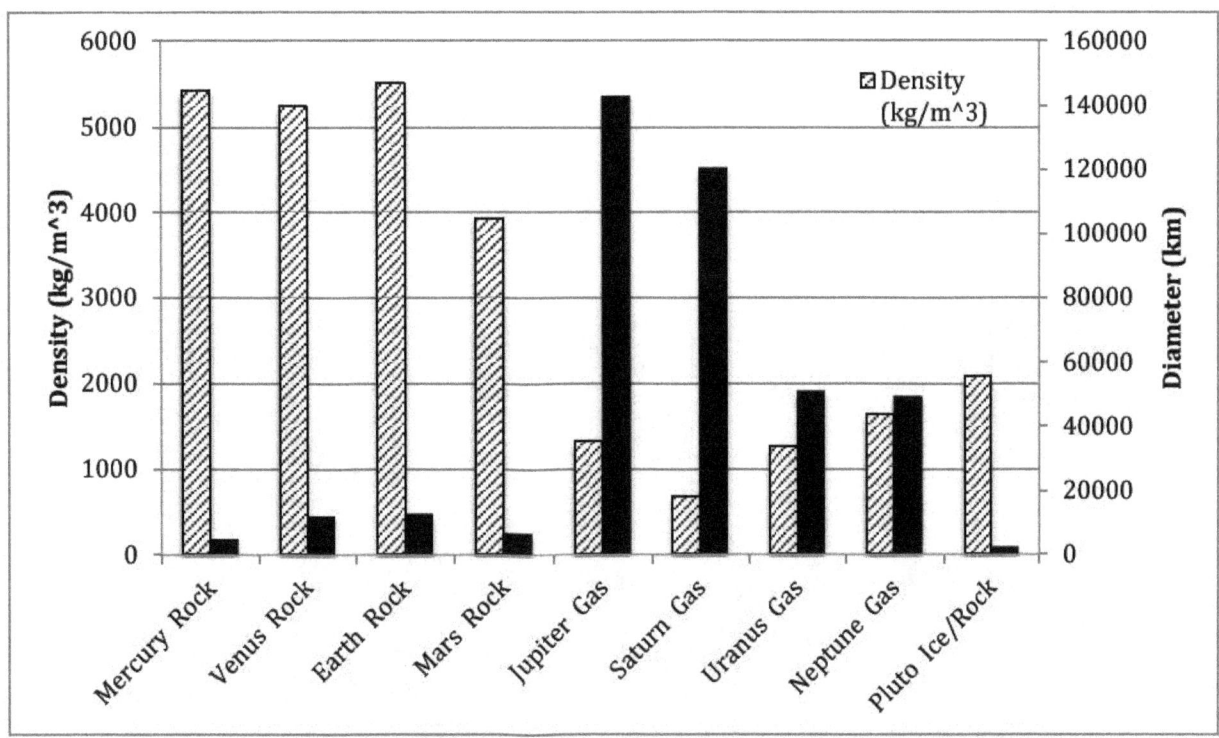

Figure 1.2: Densities, sizes, and compositions of the planets and Pluto.

What about Pluto? The International Astronomical Union (or IAU) defined a new category for Pluto: Pluto is a dwarf planet. A dwarf planet is an object that (1) orbits the Sun, (2) is big enough to be round (> 1,000 km in diameter), but (3) has NOT cleared its orbit of debris. Thus a dwarf planet satisfies the first two planet criteria but not the third.

What about objects that satisfy the first planet criteria but not the second or third? Such an object is called either an asteroid or a Kuiper Belt Object, depending upon where in the solar system it orbits. An asteroid is an object that (1) orbits the Sun between 2 and 5 AU (between Mars and Jupiter), (2) is NOT big enough to be round, and (3) is composed of rock. A Kuiper Belt Object (KBO) is an object that (1) orbits the Sun between 40 and 500 AU (beyond Neptune), (2) is NOT big enough to be round, and (3) is composed of rock and ice. Note that AU (Astronomical Unit) is a unit of distance: 1 AU = the distance between the Earth and the Sun = 93,000,000 miles. So a distance of 40 AU, for example, means 40 times further away from the Sun than the Earth is! Lastly, an object that orbits a planet or a dwarf planet is called a moon.

Figure 1.3 illustrates four expanding views of the solar system. Starting in the upper left, we have a view of the inner Solar System. Here we see the Sun and the orbits of the Terrestrial planets and Jupiter. Between the orbits of Mars and Jupiter lies the asteroid belt; asteroids are shown by dots in the image. Within the asteroid belt lies one dwarf planet (Ceres). Moving to the image in the upper right, the scale expands out to include Pluto. At this scale, the dots representing asteroids and the orbits of the inner planets are too small to pick out. However, we can see the orbits of the outer planets. Near Pluto and beyond lies the Kuiper Belt; Kuiper Belt Objects are shown by dots in the image. Additionally, there are about 50 different dwarf planets that lie within the Kuiper Belt. Some of these dwarf planets are much further away from the Sun than Pluto is. For example, the dwarf planet Sedna orbits the Sun in a very elliptical (oval) orbit that takes it as far away as about 937 AU from the Sun! The image in the lower right hand corner expands the view of the solar system to include Sedna's orbit. Lastly, surrounding the solar system there exists a spherical cloud of comets (which sometimes get bumped into the inner solar system where we can see them) called the Oort Cloud. The Oort Cloud marks the edge of the solar system and extends from about 50,000 to 200,000 AU. The image in the lower left hand corner expands the view of the solar system to include the inner edge of the Oort Cloud. Note that the Oort Cloud is so far away relative to everything else in the solar system that you can barely even see the orbit of Sedna in this image!

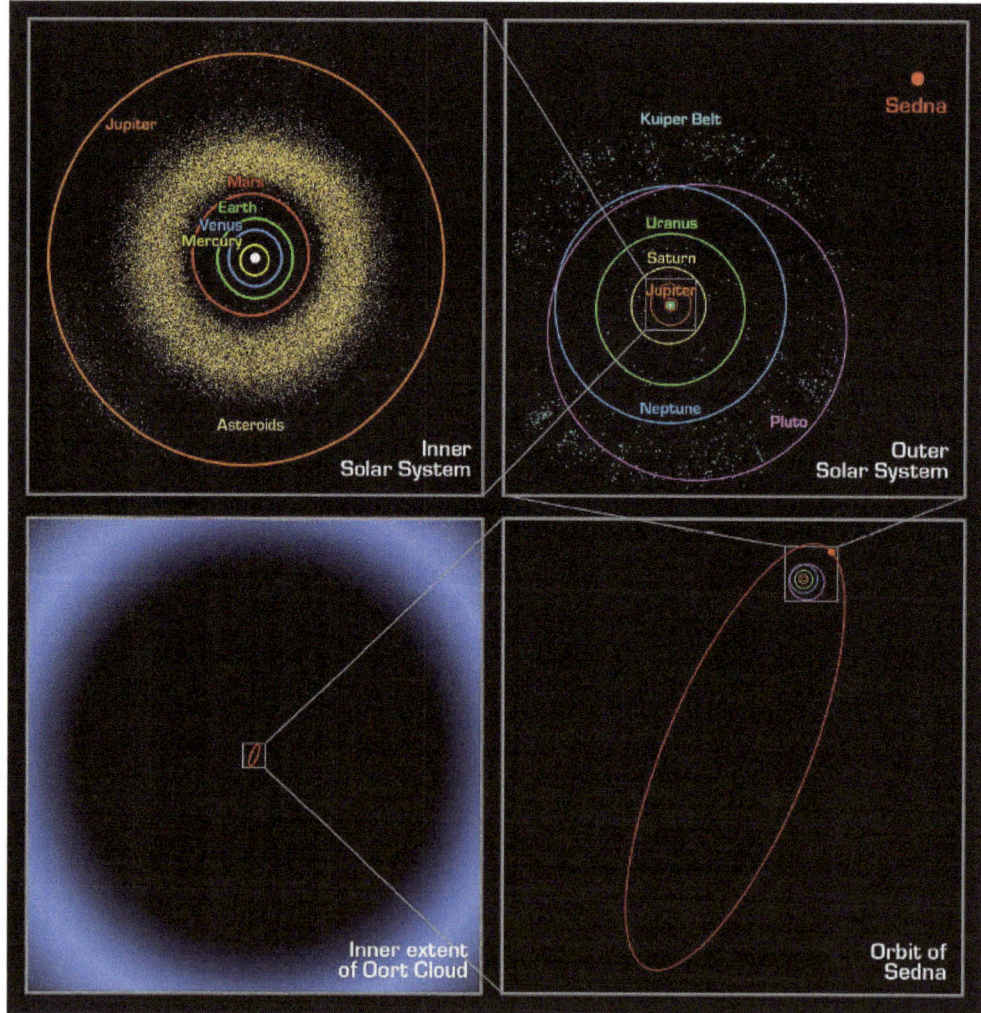

Figure 1.3: Four expanding views of the solar system, starting in the upper left and moving clockwise. (https://commons.wikimedia.org/wiki/File:Oort_cloud_Sedna_orbit.jpg)

Every chapter will end with a section called "Questions You Should Be Able To Answer." This is the basic information you should have gotten out of the chapter. I suggest that you try to answer these questions right after you finish the chapter, or while reading the chapter!

Questions You Should Be Able to Answer
1. What is a planet (according to the IAU)?
2. Why is Pluto NOT a planet? How should Pluto be classified?
3. What is the difference between a Terrestrial planet and a Jovian planet? Name the Terrestrial and Jovian planets of our solar system.
4. What is a dwarf planet?
5. What is an asteroid?
6. What is a KBO?
7. What is a moon?

Chapter 2: Solar System Sizes

In this chapter, you will gain a better understanding of the relative sizes of objects in the solar system and of their spacing.

Suppose I show you a picture of the Earth and Moon that looks like the image shown in Figure 2.1, and I ask you what is wrong with this picture?

Figure 2.1: Earth-Moon System

You probably immediately thought something like "the Earth and the Moon are the same size" or "the Earth should be bigger than the moon." And you're right. If I want to make an accurate picture of the solar system, it should maintain correct relative sizes.

If you take the diameter of the Earth and divide it by the diameter of the Moon, you get about 4. This means that 4 Moons, lined up side by side, could fit across the diameter of the Earth. So if I want to draw a correct picture, I need to make my picture of the Earth 4 times bigger than my picture of the Moon, such as is done in Figure 2.2. As another example, if I were choosing objects to represent the Earth and the Moon in a 3-D model, I would need two objects that have a size difference of 4. So for instance a 12 inch basketball and a 3 inch baseball would work.

You might now think the picture in Figure 2.2 looks correct, but it's not. That's because although the relative sizes are correct, the relative distances are not. It turns out if you take the distance between the Earth and the Moon and divide by the Earth's diameter, you get 30. This means that you can fit 30 Earths between the Earth

and the Moon. Your picture should take this into account. However, in the figure in 2.2 we can only fit about 2 and a half Earths between the Earth and the Moon! In figure 2.3, we finally have an accurate picture of the Earth and the Moon. We would call this picture a "scale model" because it maintains relative sizes and distances. The Earth is 4 times bigger than the Moon AND you can fit 30 Earths between the Earth and the Moon. As another example, if we wanted to make a 3-d scale model of the Earth and the Moon we could take the basketball and the baseball from above and separate them by 30 feet.

Figure 2.3: Accurate Scale Model of the Earth-Moon System

Shown in Figure 2.4 are the relative sizes of the Terrestrial planets. Earth is the largest terrestrial planet, Venus is just slightly smaller, Mars is about half the Earth's size, and Mercury is about a third the Earth's size. To give you a sense of the Earth's size, the diameter of the Earth is about 8,000 miles. Its circumference is about 25,000 miles. If you drove nonstop at 60 mph, it would take you about 17 days to cover that distance!

Figure 2.4: Terrestrial Planets to Scale (https://commons.wikimedia.org/wiki/File:Star-sizes.jpg)

This may sound big, but the Jovian Planets are even bigger. Shown in Figure 2.5 are the relative sizes of all of the Jovian Planets compared to the Earth. Jupiter's diameter is about 11 times bigger than the Earth's; meaning that 11 Earths lined up side by side would fit across Jupiter. Saturn is slightly smaller than Jupiter. Lastly, Uranus and Neptune are both about 4 times larger in diameter than the Earth.

Figure 2.5: Earth and Jovian Planets to Scale (https://commons.wikimedia.org/wiki/File:Star-sizes.jpg)

However, none of the planets compare in size to the Sun (Figure 2.6). If we add the Sun to our picture, we see that it's about 110 times larger in diameter than the Earth; in fact, you can hardly see the Earth in this picture. Looking at this picture, it should be unsurprising that the Sun contains 99.86% of all of the mass in the solar system. In terms of volume, you could fit 1.3 million Earths inside the Sun!

Figure 2.6: Planets and Sun to Scale
(https://commons.wikimedia.org/wiki/File:Planets_and_sun_size_comparison.jpg)

What about distances? We often think of the planets as being evenly spaced from the Sun, but that is not that case. In fact the spacing of the planets is such that it is difficult to see all of the planets' orbits in one diagram. Figure 2.7 shows the planets' orbits to scale. The left side of the figure shows the planets' orbits out to Mars. The right side of the figure shows the planets' orbits from Mars to Pluto. All of the terrestrial planet's orbits would fit within the circle representing Mars' orbit.

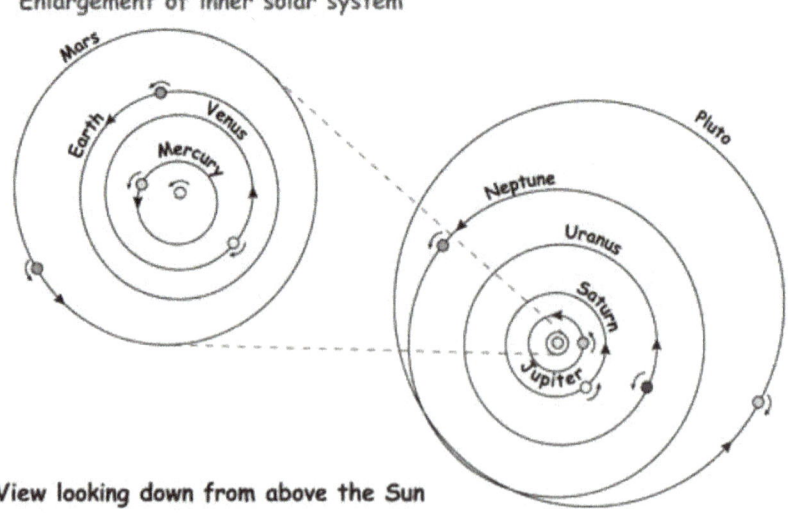

Figure 2.7: Planet Orbits to Scale (http://www.nasa.gov)

Table 2.1 shows the actual distances of the planets from the Sun and their diameters. The distances are given in AU. An AU is an astronomical unit. 1 AU is equal to the distance between the Earth and the Sun, which is 93 million miles. So since Uranus has a distance of 19 AU, this means that it is 19 times further from the Sun than the Earth is.

Planet Name	Distance (AU)	Diameter (km)
Mercury	0.387	4,879
Venus	0.723	12,104
Earth	1.000	12,742
Mars	1.524	6,779
Jupiter	5.203	139,822
Saturn	9.537	116,464
Uranus	19.191	50,724
Neptune	30.069	49,244

Table 1.1: Distances and Diameters of the Planets

Questions You Should Be Able to Answer
1. What is a scale model?
2. You should have a general idea of the sizes and distances in the solar system.

Suggestions for Further Reading
1. http://www.scalesolarsystem.66ghz.com/#saturn (scale model example)
2. http://thinkzone.wlonk.com/SS/SolarSystemModel.php (scale model calculator and map)

Chapter 3: Studying the Planets

In this chapter, you will learn how scientists study objects in the solar system. You will learn how science is defined, how the scientific method works, and the difference between a hypothesis and a theory. Then you will learn how scientists apply the scientific method to learn about the planets, and the different types of robotic space missions they use.

Science is hard to define because it is both knowledge and a process. By knowledge, I mean that science is what we know about the world; in other words, all of the things you read about in your science textbooks: the general principles that operate in the world, as we have discovered through observation and experimentation. Things like Newton's Laws and the Theory of Evolution. By process, I mean that science is the process we use to obtain this knowledge; in other words, the scientific method.

An alternative way to attempt to define science is to look at its characteristics (see Figure 3.1). The first characteristic is that science focuses on the natural world (things like atoms, plants, galaxies, and the force of gravity). Unobservable supernatural phenomenon, like ghosts, are not allowed because they don't operate according to the known rules that govern the natural world and therefore can't be studied. Not only does science focus on the natural world, but it aims to explain that world. For instance, science tries to answer questions like why we have earthquakes and why the planets all look different from one another. The third characteristic is that science uses testable ideas. For an idea to be testable, it must generate specific predictions that would be born out if the idea were true (we will discuss this more later). Next, science relies on evidence. This means that ideas in science have data to back them up. Fifth, science involves a community of scientists working together to come to a consensus – it is not just one person working alone. A lot of science is done in collaborations, with groups of scientists working together. Furthermore, a scientific idea is not considered useful until it has been presented to, validated by, and used by other scientists. Next, science leads to ongoing research. This means it will never be finished – there will always be more ideas to explore and to test. And lastly, science benefits from scientific behavior. This means that science works best when scientists behave in an ethical manner; they shouldn't make up data, steal other's work, and so on.

Science...

- focuses on the natural world
- aims to explain the natural world
- uses testable ideas
- relies on evidence
- involves a community
- Leads to ongoing research
- benefits from ethical behavior

Figure 3.1: Science Characteristics

So what can science do? Well, science seeks to explain and make predictions about the natural world. However, it is important to understand that science will never uncover a 100% correct and finished model of the world. This is not possible because all science is tentative. There is always another test that you can do, and there is always a chance that that test will change your ideas. In short, we can say that science gives us a model of nature that is accurate (meaning it fits our observations) and is useful (meaning we can use the model to make reliable predictions). For example, Newton's Laws are accurate in that they fit all of our observations about the motions of objects, and they are useful in that we can use them to make reliable predictions, like how much fuel you need to send a rocket to the moon!

You have probably seen the scientific method represented as a linear series of steps in the past (such as in Figure 3.2). However, this isn't really an accurate illustration of how science works. All of these steps, like constructing a hypothesis and doing experiments, are important parts of science, but science is not done in any particular order and there are lots of dead ends.

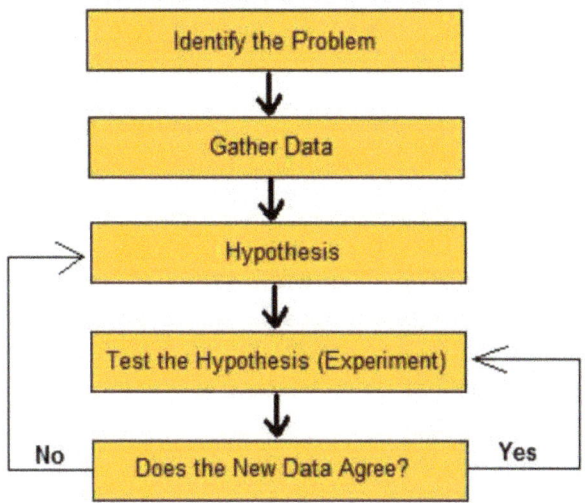

Figure 3.2: Too Simple A View of the Scientific Method (https://commons.wikimedia.org/wiki/File:The_Scientific_Method.jpg)

It is better instead to represent the scientific method in a diagram as shown in Figure 3.3, which shows all of the parts involved, but in no specific order. At the heart of science is testing ideas. This is what science is all about: gathering and interpreting data to test the validity of an idea (note that in science, ideas are known as hypotheses). At the periphery of the diagram, we have several other important parts of science. At the top is exploration and discovery – this is where scientists observe, read, ask questions, and in general search for inspiration. In the bottom right is community analysis and feedback. This is the community endeavor part of science, where a scientist gets feedback from her peers through conferences, publications, and discussions. Lastly in the bottom left is benefits and outcomes. This is why we do science in the first place. To develop new technology, solve problems, address, societal issues, inform policy, and build a new knowledge.

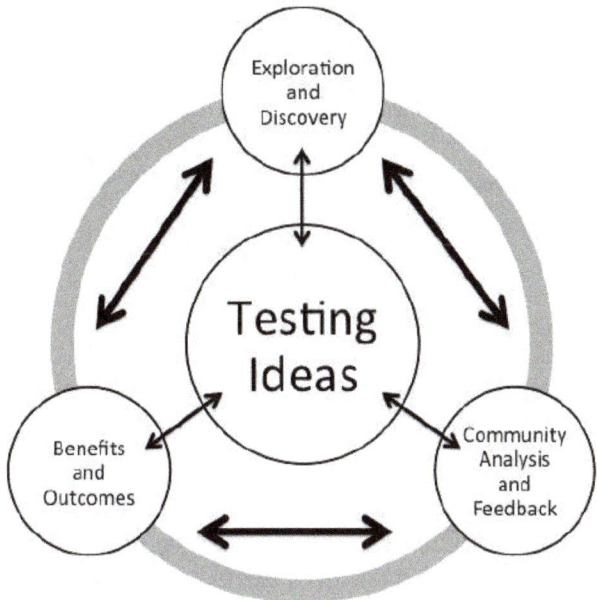

Figure 3.3: Scientific Method (modified after http://undsci.berkeley.edu/article/intro_01)

Focusing on hypotheses, a hypothesis is an explanation of a phenomenon. A good hypothesis should be both testable and evidence-based. Testable means that a hypothesis should make a prediction that can be tested through experiments or gathering new data. For example, one could make a hypothesis that plants need water to grow. You could test this via an experiment – try to grow plants without water and see what happens. Of course, you'll find the plants will not grow, and your hypothesis will be supported. An example of a hypothesis that would not be testable is that ghosts help plants to grow. This is not testable because (despite what you may see on TV) there is no way to test for the presence of ghosts! Evidence-based means that it is supported by the majority of the current data we've collected. Note that a hypothesis could be evidence-based at one point in time, but could be rejected later if new data is collected that conflicts with the hypothesis' predictions.

Many people confuse hypotheses and theories. But these concepts have very specific and different meanings in science. In science, a hypothesis is just an idea. You can call something a hypothesis whether or not it has been tested yet. A theory, on the other hand, is a model of some piece of nature that has been tested many many times and has passed all of the tests so far. Theories therefore have tons of evidence to support them. When scientists say that evolution is a theory, what we mean is that there is a lot of evidence and experiments in support of that idea. Additionally, theories tend to be broader in nature than hypotheses; they explain a wide range of phenomena.

Going back to the scientific method, let's focus on gathering and interpreting data to test a hypothesis. How do we do this when the objects we are studying are millions of miles away, like the planets?

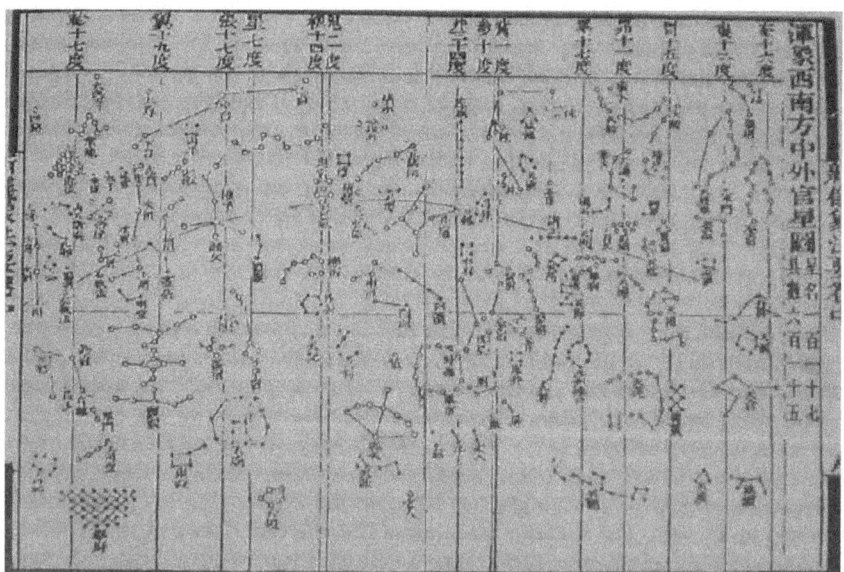

The simplest way to go about this is naked eye observation. Go out at night, look at the sky, and track the motions of objects over the course of the night, the week, the month, the year, and so forth. Astronomers have been doing this for centuries. For example, the Mayans built their temples around celestial events. They would design their temples so that the stars and the Sun would appear in certain places around the temple at special times of the year. Additionally, the Chinese developed very detailed star maps (Figure 3.4), and the Greeks developed sophisticated models of the solar system.

Figure 3.4: Ancient Chinese Star Map
(https://commons.wikimedia.org/wiki/File:Su_Song_Star_Map_1.JPG)

To learn more, one needs telescopes to see objects that are too dim to be visible to the naked eye. The first telescopes were developed in the early 1600s and were little more than giant spyglasses. Today's telescopes are highly advanced and allow us to see even to distant galaxies.

Scientists also study the solar system using experiments. As an example of the type of experiment that scientists use to study the solar system, Figure 3.5 shows a picture of the Shock Compression Lab at Harvard University. Here scientists use a giant gun to smash pieces of rock against one another to simulate what happens during meteorite impacts. Sometimes problems are so broad in scale that scientists need to do experiments on the computer; Figure 3.6 shows an example of a simulation that a scientist would use to study what happens when two asteroids collide.

An additional way that scientists can study the solar system is using meteorites. Meteorites are rocks that fall to Earth from space. Most meteorites come from the asteroid belt. However, some asteroids actually come from other places, like Mars or the Earth's Moon!

While naked eye observations, telescopes, experiments, and meteorites give us some information about the solar system, MOST of our information about the solar system comes from space exploration: both manned space missions and robotic space missions. To date, three countries have sent humans into space: The Soviet Union, The United States, and China.

The US manned space program can be broken down into 6 main programs. The Mercury program, from 1959-1963, put humans into space and into orbit around the Earth. The Gemini program, from 1963-1966, practiced skills (like spacewalks) that would be needed to send astronauts to the moon. The Apollo program, from 1961-1972, put humans on the moon. Skylab, from 1973-1974, was the first American space station. The Space Shuttle was a reusable space vehicle which we used from 1981-2011. Lastly, the International Space Station is a large research space station currently in use. Astronauts used to be shuttled back and forth to the International Space Station via the Space Shuttles, but now a Russian vehicle, called Soyuz, is used instead.

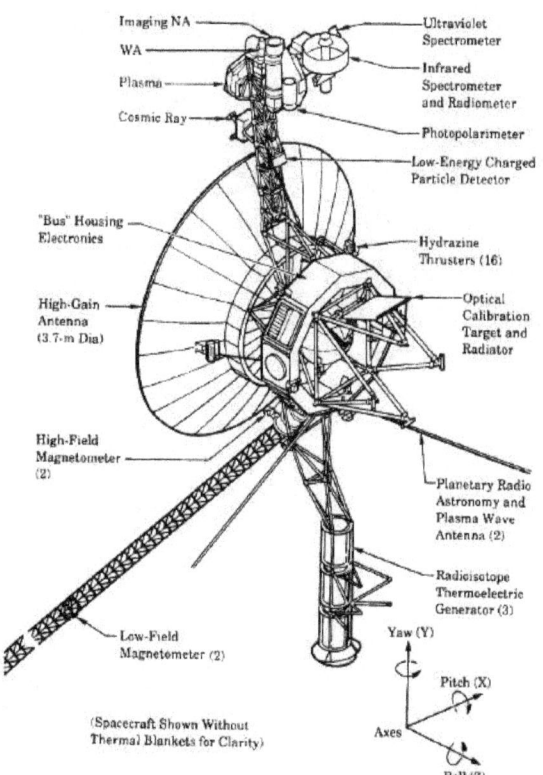

Figure 3.7: The Voyager Spacecraft
(https://commons.wikimedia.org/wiki/File:Voyager)
_spacecraft_structure.jpg

The pinnacle of the manned American space program was Apollo. Since then, manned space program has dropped off in favor of robotic space missions. The simple reason for this is funding: NASA's budget dropped sharply following Apollo (NASA now receives less than 0.5% of the federal budget), and robotic space missions are much cheaper than manned space missions.

There are six kinds of robotic space missions that we've used to explore the solar system: they are flybys, orbiters, landers, rovers, sample returns, and impactors. We will go through each type in turn.

Before we do so, however, let's look at the kinds of instruments spacecraft can carry. They can carry cameras of various resolutions to take pictures. They can also carry spectrometers, which are instruments that measure the kinds and amounts of light coming off of a surface. This allows us to determine compositions (like the minerals and rocks that make up a surface, or the gases in an atmosphere) because different substances have different spectroscopic signatures, analogous to the way different people have different fingerprints. Altimeters measure topography, magnetometers measure the size and direction of the magnetic field, seismometers measure motions of the ground (due to earthquakes), thermometers measure temperature, anemometers measure wind speed, and barometers measure atmospheric pressures. Lastly, all spacecraft will have radio transmitters and antennas to communicate with the Earth.

A flyby mission flies by an object (or several objects) and collects data along the way. Because the amount of data collected is limited, this kind of mission is good for reconnaissance. For example, suppose you want to explore an area of the solar system that has never been explored before and to take a little bit of data on a lot of objects to determine what is out there and what might be interesting for more detailed future study....a flyby would be ideal for this. Examples of flyby missions are the Voyager I and II missions (Figure 3.7), which were launched in 1977. These missions were the first to travel to the outer solar system and performed flybys of the Jovian planets and their satellites. A fun bit of trivia is that when scientists launched the Voyager missions, they recognized that the spacecraft would eventually drift away from our solar system and one day, tens of thousands of years from now, perhaps enter a different solar system and just maybe be found by intelligent alien life! Just in case this ever did happen, scientists included on each spacecraft a gold-plated phonograph record containing sounds and images selected to represent the diversity of life on Earth. The cover of the record contains images to help potential alien life decipher where the spaceship is from and to play the record. A sample of the statement left on the record by then-President Jimmy Carter reads: "This is a present from a small distant world, a token of our sounds, our science, our images, our music, our thoughts, and our feelings. We are attempting to survive our time so we may live into yours. We hope someday, having solved the problems we face, to join a community of galactic civilizations."

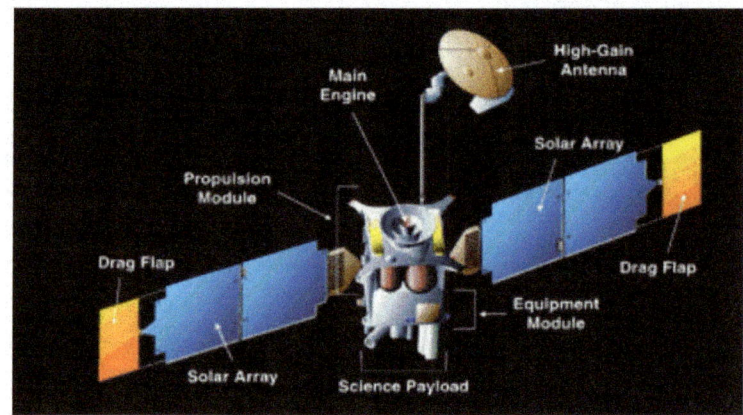

Figure 3.8: Mars Global Surveyor (http://www.nasa.gov)

Now suppose you want more comprehensive data. Not just a few pictures of a planet, but maps of the entire surface. Then you would use an orbiter. Orbiters are put into orbit around a planet to collects lots of data over a long period of time. Not only does this give you a view of the entire planet, but it allows you to see any changes that may occur over time. An example of an orbiter mission is the Mars Global Surveyor, which was launch in 1996 and gave us a detailed view of Martian climate and geology.

For some measurements, you need to be actually on the surface. For example, you cannot measure wind speed or analyze a planet's soil unless you are sitting on the surface. Thus the next type of robotic space mission is a lander. Landers land on a planet but stay put once they land. This allows us to get an up-close detailed view of the surface. An example of a lander is Viking 1, which was the first lander on Mars (Figure 3.9).

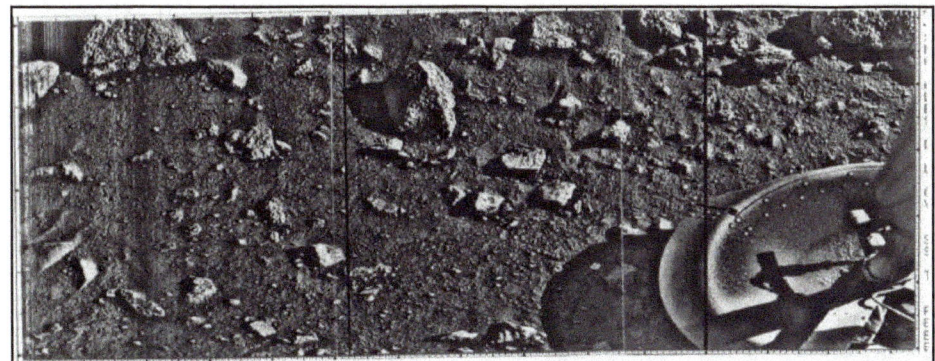

Figure 3.9: First Image Ever Taken from the Surface of Mars, taken by Viking 1 on July 20, 1976
(http://www.nasa.gov/)

Sometimes you not only want to land, but you want to move around as well. For this you would use a rover. Not only does this give you an up-close detailed view of the surface, but you can move to interesting things! Rovers typically move about 10 miles in a period of few years. An example is the Mars Exploration Rovers, Spirit and Opportunity (Figure 3.10), which were launched in 2003.

Figure 3.10: Mars Exploration Rovers (http://www.nasa.gov/)

The goal of a sample return mission is to bring stuff (like rocks or dust) back so that scientists can study it in labs here on Earth. So far, there have not been any sample return missions to other planets (due to cost - you have to carry enough fuel to not only launch off of the Earth, but the destination planet as well); however, Stardust (Figure 3.11), which was launched in 1999, brought back comet dust.

Figure 3.11: Stardust (http://www.nasa.gov/)

Last, but not least is an impactor mission. In this type of mission, you have an object hit a surface and create an explosion. This allows you to see what lies underneath the surface. An example of this type of mission is the Deep Impact mission to comet Tempel 1 (Figure 3.12). This mission launched an impactor into the comet whilst a flyby collected data. This allowed scientists to determine the structure of the comet and figure out the comets are essentially dirty snowballs!

Figure 3.12: Impact Created by Deep Impact (http://www.nasa.gov/)

Questions You Should Be Able to Answer
1. How do we define science?
2. What are some of the important characteristics of science?
3. What can science do?
4. What can science <u>not</u> do?
5. Describe the scientific method. What is at the heart of the scientific method?
6. What is a hypothesis?

7. What makes a good hypothesis?
8. What is the difference between a hypothesis and a theory?
9. What are some ways to gather data bout the solar system?
10. Describe a flyby, orbiter, lander, rover, sample return, and impactor mission. What is each type of mission best suited for?

Suggestions for Further Reading
1. http://undsci.berkeley.edu/article/intro_01 (the scientific method)
2. http://www.nationalgeographic.com/125/timelines/space-exploration/ (space exploration timeline)

Chapter 4: The Copernican Revolution

In this chapter, you will learn how scientists developed an accurate model of the solar system. You will see the assumptions that were made by ancient astronomers and how a phenomena called retrograde motion challenged these assumptions. You will then see how Ptolemy and Copernicus explained retrograde motion using different models of the solar system.

You are probably aware that every day the Sun rises in the east, moves across the sky, and sets in the west. Figure 4.1 shows the path of the Sun over the course of a day; note that in this diagram you are looking south, so east is to the left of the page and west is to the right. At the beginning of the day, the Sun will rise in the east. It will move higher in the sky until it reaches its highest point in the southern sky at noon. Then it will descend and finally set in the west. Note that this is what they sun *appears* to do, but in *reality* the sun is not actually moving. The reason for the apparent motion is that the Earth rotates. The Sun itself doesn't move, but as the Earth rotates, the Sun will appear to move across the sky.

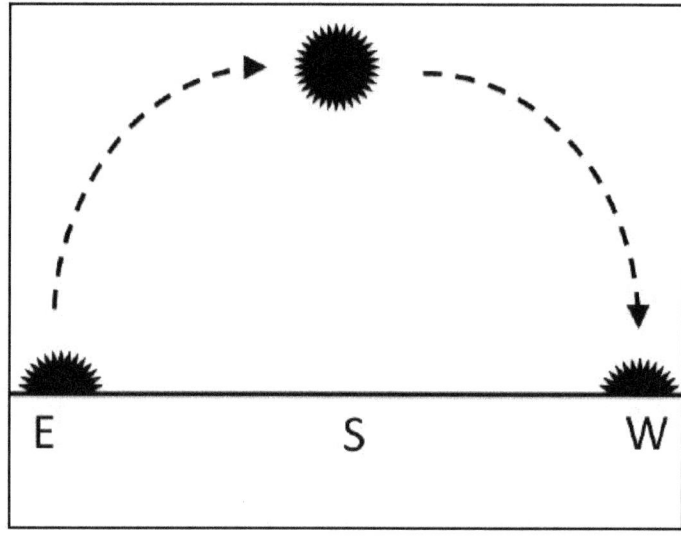

Figure 4.1: Path of Sun in Sky Over a Day

By the same logic, any other object in the sky that doesn't move over the course of a day will exhibit the same apparent motion. So not only does the Sun appear to rise in the east and set in the west, but every other object in the sky – like stars and planets – do so as well.

Of course ancient astronomers did not know this. Because we see things appear to move around the Earth, ancient astronomers believed that the Earth is the center of the solar system and that the Earth doesn't move.

However, this assumption leads to problem when trying to explain the way in which planets appear to move in the sky. Over the course of a single night, a planet will move with the stars, rising in the east and setting in west. However, over the course of weeks or months, the planet will move *relative* to the background stars. So though the planet will always move with the stars on any given day, it will be next to different stars as the weeks progress. Another way to think about this is if you take a picture of the sky at midnight one day, and then take a picture of the sky at midnight one week later, the stars will all be in roughly the same position and in the same patterns, but the planet will have moved.

Figure 4.2 shows the motion of Mars relative to the background stars during the summer of 2018. The image is looking south, so east is the left and west is to the right. Note that in general planets appear to drift west to east relative to the background stars. This is known as prograde motion. However, sometimes a planet will temporarily reverse direction and move from east to west. This is known as retrograde motion. Figure 4.3 shows the same picture as Figure 4.2, but the prograde motion is highlighted in light gray and the retrograde motion is highlighted in dark gray.

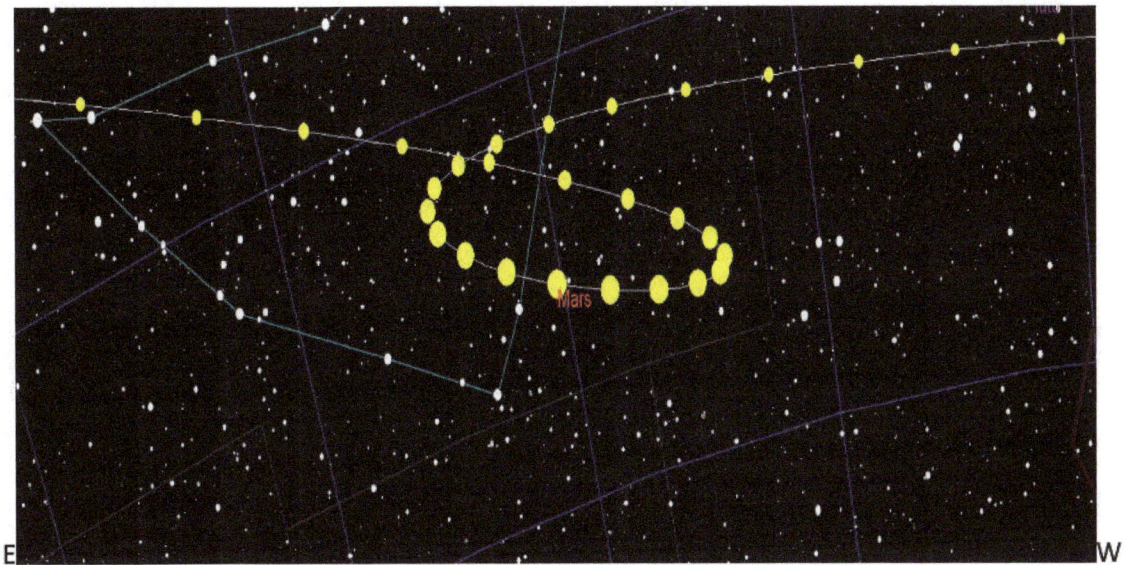

Figure 4.2: Motion of Mars Relative to Background Stars
(https://commons.wikimedia.org/wiki/File:Mars_motion_2018.png)

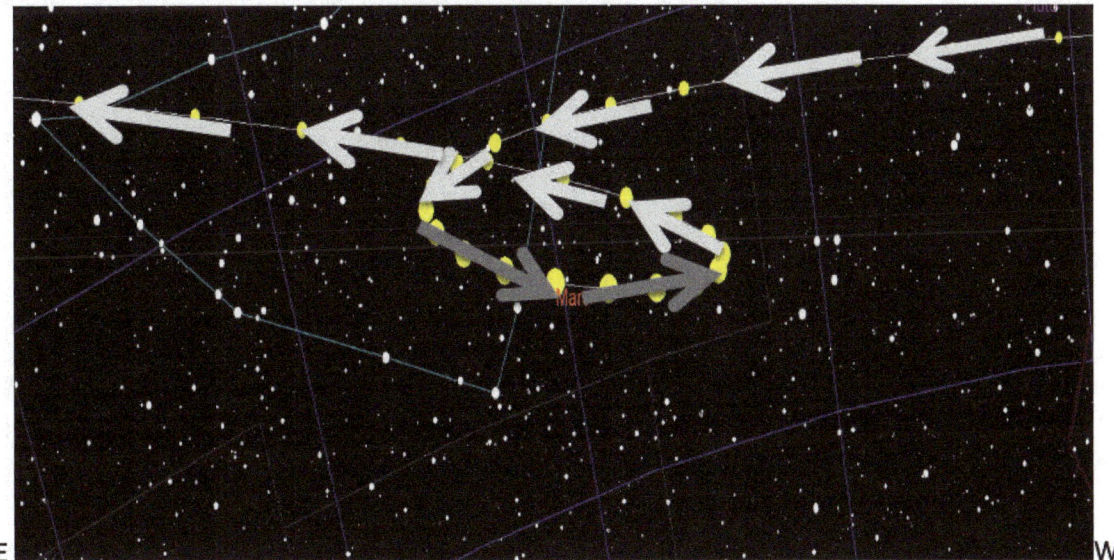

Figure 4.3: Motion of Mars Relative to Background Stars. Prograde motion is highlighted in light gray, and retrograde motion is highlighted in dark gray.

Retrograde motion cannot be explained by a model of the solar system where the planets orbit directly around the Earth. So in AD 140, the Greek astronomer Ptolemy developed a model of the solar system to explain this motion (Figure 4.4). This was the first mathematical model of the solar system. In this model, the Earth is at the center, and the sun and planets orbit around the Earth. However, the planets do not orbit *directly* around the Earth. Instead, they orbit around small circles (called epicycles) and the epicycles orbit the Earth. Because of this simultaneous rotation, the planet will sometimes move backwards in the sky (hence, retrograde motion). It should also be noted that Ptolemy's model didn't technically have the Earth directly at the center of the planets' orbits; he had a point slightly offset from the Earth that he called the "Center of the Deferent" at the center.

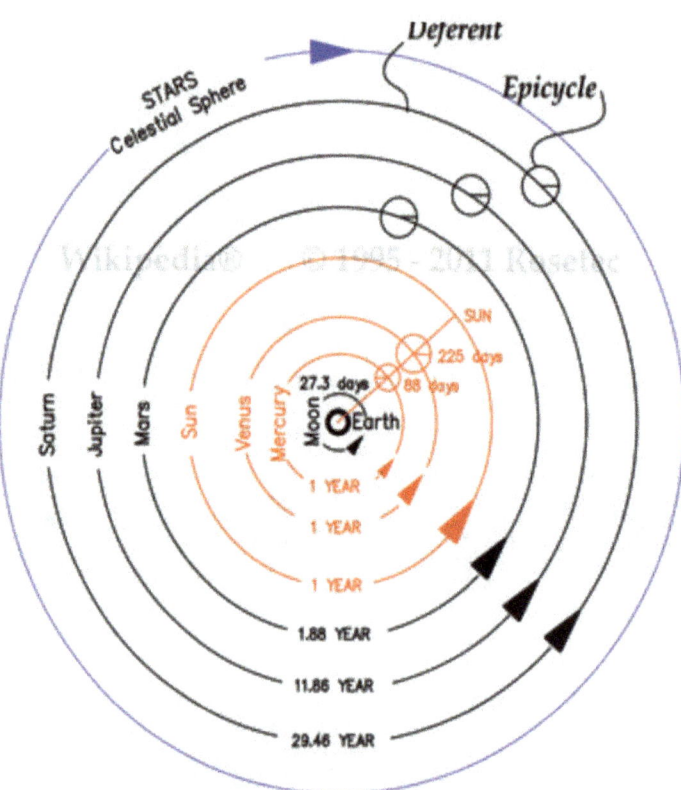

Figure 4.4: Ptolemy's Model of the Solar System
(https://upload.wikimedia.org/wikipedia/commons/8/82/De_revolutionibus-
Copernicus_Illustrates_Heliocentric_Order_of_Planets_not_Orbits.jpg)

While Ptolemy's model explained retrograde motion (Figure 4.5), it was a complicated model and became even more complicated. As data on the solar system improved, Ptolemy (and others) found that they had to add more and more epicycles to explain the data. For instance, instead of a planet orbiting on an epicycle, you have a planet orbiting an epicycle, which orbits on another epicycle, which orbits around the Earth. Late versions of Ptolemy's model had nearly 100 layered epicycles to explain the observations!

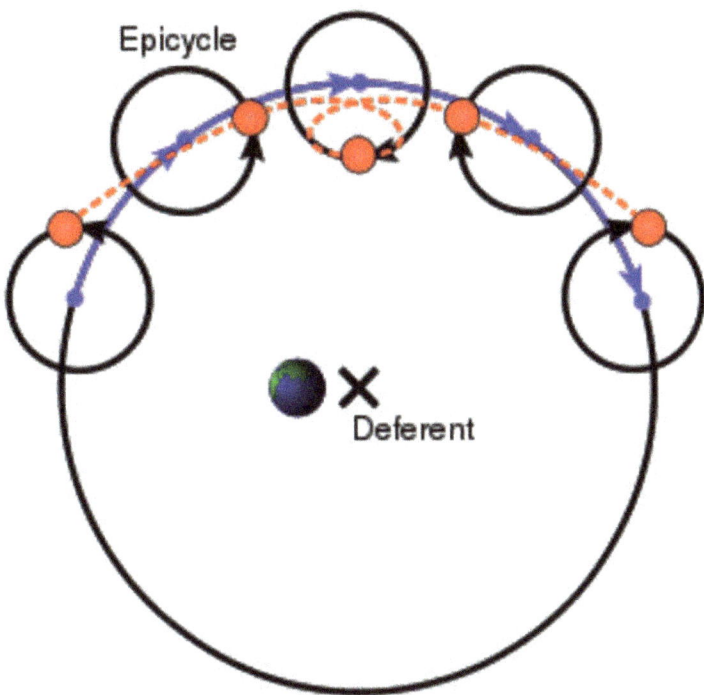

Figure 4.5: Ptolemy's Model Explains Retrograde Motion (apparent path of planet is dashed line)
(https://commons.wikimedia.org/wiki/File:Ptolemy_Epicycles.svg)

In the early 1500s, Copernicus realized that things would be a lot simpler if the Sun were at the center of the solar system. If you let the Sun be at the center, and let the planets that are closer orbit faster than the planets that are further, then retrograde motion results as a natural consequence.

This is illustrated in Figure 4.7. We have the Sun at the center (labeled S), and the orbits of Earth (inner circle) and Mars (outer circle) are shown. The position of the Earth (the numbered Ts) is shown at seven different times, as well as the corresponding positions of Mars (the numbered Ms), and the corresponding position of Mars relative to the background stars as seen from Earth (the numbers). Notice that Mars in general drifts towards the east relative to the background stars (prograde motion). However, because the Earth is moving faster, it eventually overtakes Mars in its orbit. When this happens Mars will appear as if it is moving backwards (retrograde motion). Note that it is not actually moving backwards; it just looks that way. This is just like when you are in a car and you approach and overtake a slower car on your right. If you look over at the slower car at the moment you pass, it will look to you as if that car is moving backwards (try it next time you're in the car and not driving!). The car is not actually moving backwards, it just looks that way. In Figure 4.7, Mars is in prograde motion from dots 1 to 3, retrograde motion from dots 3 to 5, and prograde motion again from dots 5 to 7.

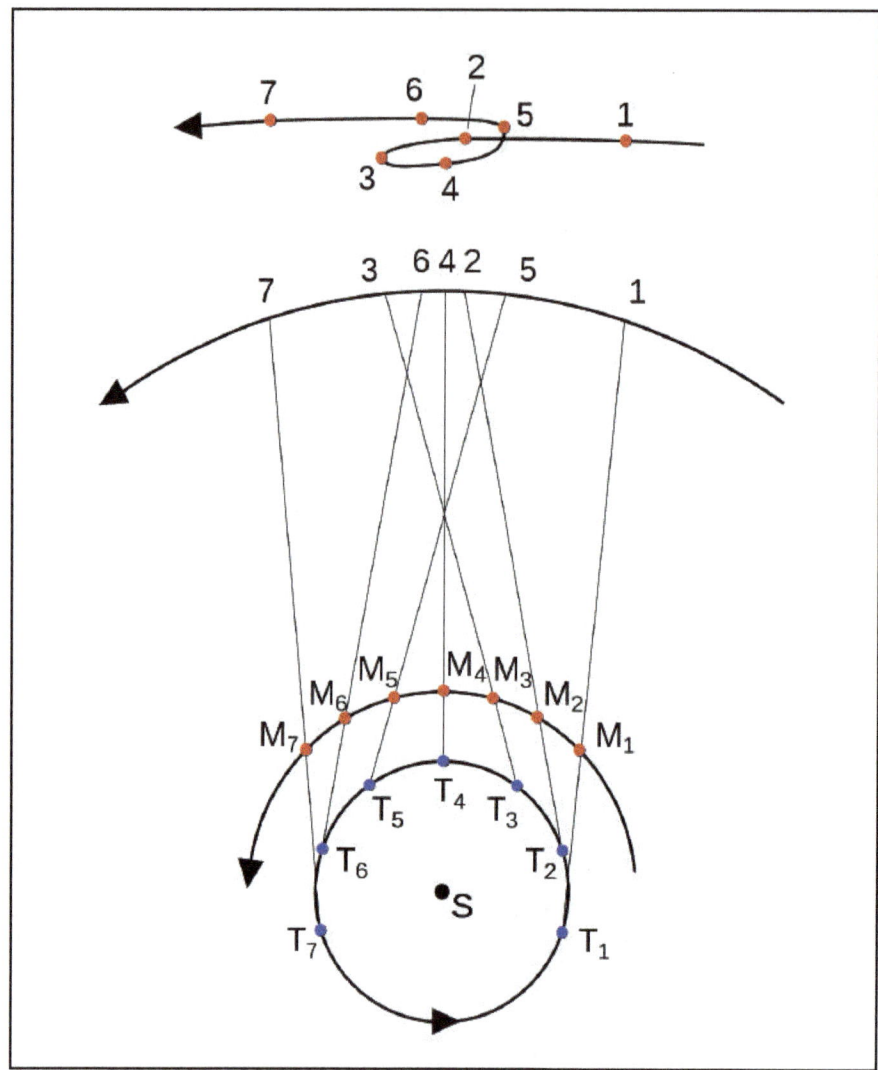

Figure 4.7: Copernicus' Model of the Solar System Explains Retrograde Motion
(https://commons.wikimedia.org/wiki/File:Qualitative_apparent_retrograd_motion_of_Mars_top-down.svg)

In science, when there is more than one hypothesis that correctly explains the data, the one that is simpler and makes less assumptions often turns out to be the correct one (this is known as Occam's Razor). This example is no exception; though it took some time for the Earth-centered model to lose its hold, Copernicus' sun centered model eventually won out!

Questions You Should Be Able to Answer
1. Why did ancient astronomers believe that the Earth was the center of the solar system and that the Earth doesn't move?
2. What do we mean by the "retrograde motion" of the planets?
3. Describe Ptolemy's model of the solar system. How did this model explain retrograde motion?
4. Describe Copernicus' model of the solar system. How did this model explain retrograde motion?

Suggestions for Further Reading
1. http://www.lasalle.edu/~smithsc/Astronomy/retrograd.html (explanation of retrograde motion)

2. http://astro.unl.edu/classaction/animations/renaissance/marsorbit.html (Ptolemy's model retrograde motion animation)
3. http://astro.unl.edu/classaction/animations/renaissance/retrograde.html (Copernicus' model retrograde motion animation)

Chapter 5: Kepler's Laws

In this chapter, we will continue to discuss how scientists developed an accurate model of the solar system. You will learn about Kepler's three laws of motion, which can be used to accurately describe planetary motions.

In the mid to late 1500s, a gentleman named Tycho Brahe set out to create the most mathematically accurate model of the solar system to date. To help him, he hired an assistant, Johannes Kepler. Unfortunately, Brahe held strongly to an Earth-centered view of the solar system, and began to create his model in the style of Ptolemy, with many epicycles. However, Brahe died before the model was complete, leaving Kepler to finish. Kepler then abandoned the Earth-centered view in favor of the Sun-centered view. In doing so, Kepler realized that three laws can describe the motions of the planets. We now call these laws "Kepler's Laws of Planetary Motion."

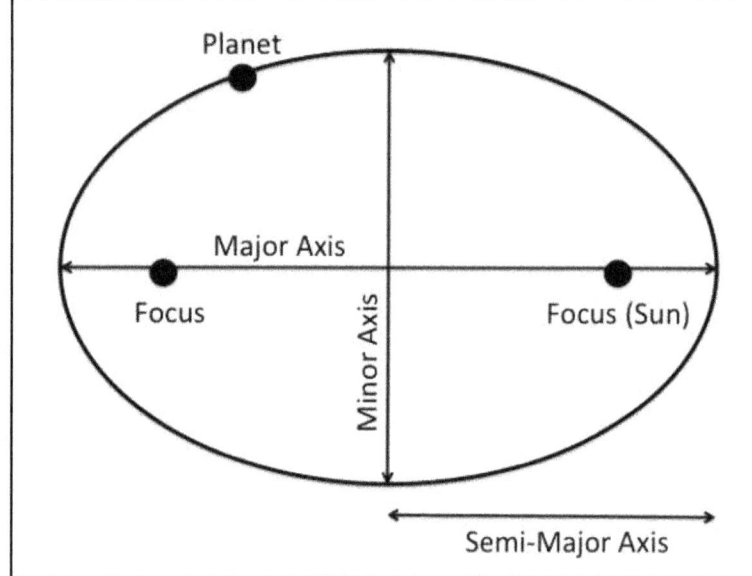

Figure 5.1: Ellipse Terminology

Kepler's 1st Law states that the planets orbit the Sun in ellipses with the Sun at one focus. This means that the planets' orbits are not perfectly circular; they are in fact slightly oval. Figure 5.1 illustrates some important terminology. Just as a circle has a center, an ellipse has two foci. The Sun is at one focus; the other focus is empty. The long axis of the ellipse is known as the major axis, and the short axis is known as the minor axis. Half of the major axis is the semi-major axis. This is important because it is mathematically equivalent to the planet's average distance from the Sun. Note that any point on the ellipse has some distance to the Sun and some distance to the empty focus. An ellipse is defined so that that sum of these distances is always the same, no matter where on the ellipse you are. An ellipse can be more oval or less oval; we describe the degree of ovalness with the eccentricity, which is a decimal in between 0 and 1. An eccentricity of 0 represents a perfect circle, and an eccentricity of 1 represents a straight line. Figure 5.2 shows examples of ellipses with different eccentricities. When the eccentricity is 0, the two foci are at the same place and we have a perfect

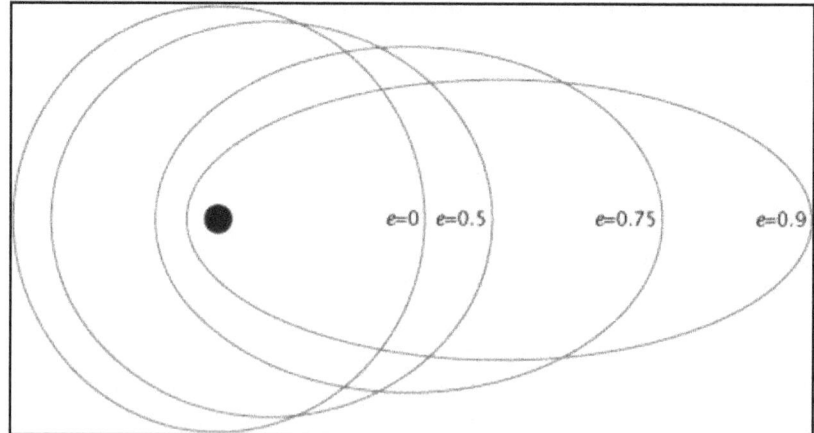

Figure 5.2: Ellipses With Different Eccentricities, Same Semi-Major Axis (http://nasa.gov)

circle. As the eccentricity increase, the foci move further and further apart, and the ellipse becomes more and more oval.

Kepler's 2nd Law states that a line from a planet to the Sun sweeps over equal areas in equal intervals of time. The example in Figure 5.3 illustrates this. Note that Area A_1 = Area A_2 = Area A_3; the amount of shading is the same. Thus the planet takes the same amount of time to cover the distance represented by Area A_1, the distance represented by Area A_2 and so forth. Note also that though the areas are the same, *the distances are not the same*! The distance closer to the Sun (e.g.the distance corresponding to Area A_3), is in fact bigger than the distance further from the Sun (e.g. the distance corresponding to Area A_1). This means that when the planet is closer to the Sun, it has to cover a greater distance in the same amount of time. In order to do this, it must be traveling faster.

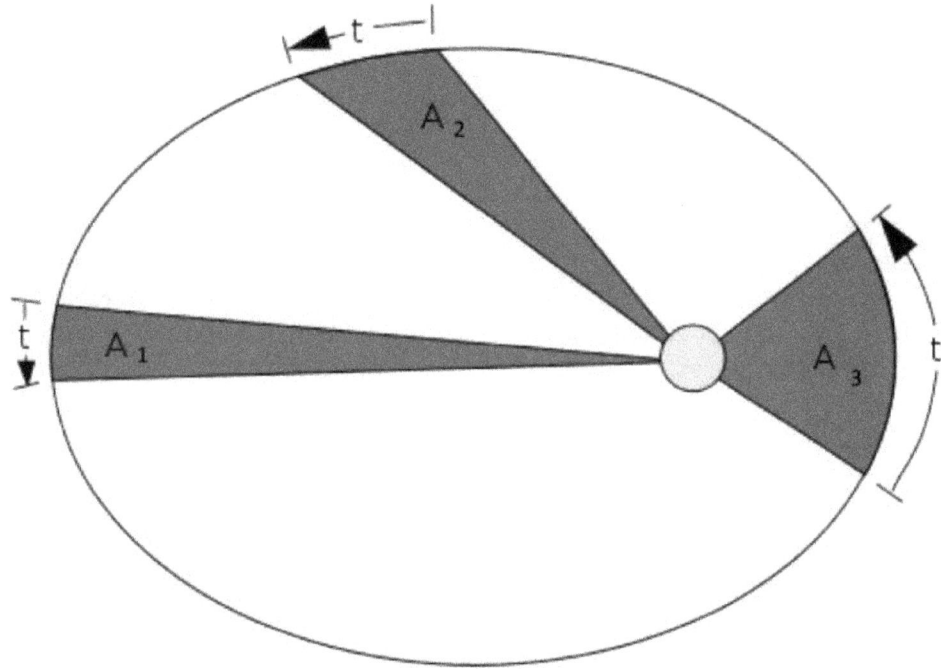

Figure 5.3: Kepler's 2nd Law Illustration
(https://commons.wikimedia.org/wiki/File:Kepler%27s_law_2_ru.svg)

Thus a consequence of Kepler's 2nd Law is that the speed a planet travels during its orbit changes and is related to its distance from the Sun. When the planet is closer to the Sun it travels faster, and when the planet is further from the Sun, it travels slower. Thus the planet's slowest speed will be at the furthest point from the Sun, and the fastest speed will be at the closest point to Sun (Figure 5.4). As the planet gets closer to the Sun it will speed up, and as it gets further from the Sun it will slow down. So in diagram 5.4 if the planet is traveling counterclockwise, all along the top half of the orbit shown it will be slowing down (at the points on the orbit marked by Xs) and all along the bottom half of the orbit shown it will be speeding up (at the points on the orbit marked by asterisks). Note: be careful not to confuse how fast the planet is moving with whether or not it is speeding up or slowing down! The speed is determined by the distance; whether it is speeding up or slowing down is determined by whether it is currently moving away or towards the Sun. For example, consider the X closest to the fastest point and the asterisk closest to the fastest point. Both are the same distance from the Sun, so at both points the planet is traveling the same speed (which is pretty fast since it is close to the Sun). However, at the asterisk the planet is speeding up (think of a car going 60 mph where you are stepping on the gas) and at the X the planet is slowing down (think of a car going 60 mph where you are stepping on the breaks).

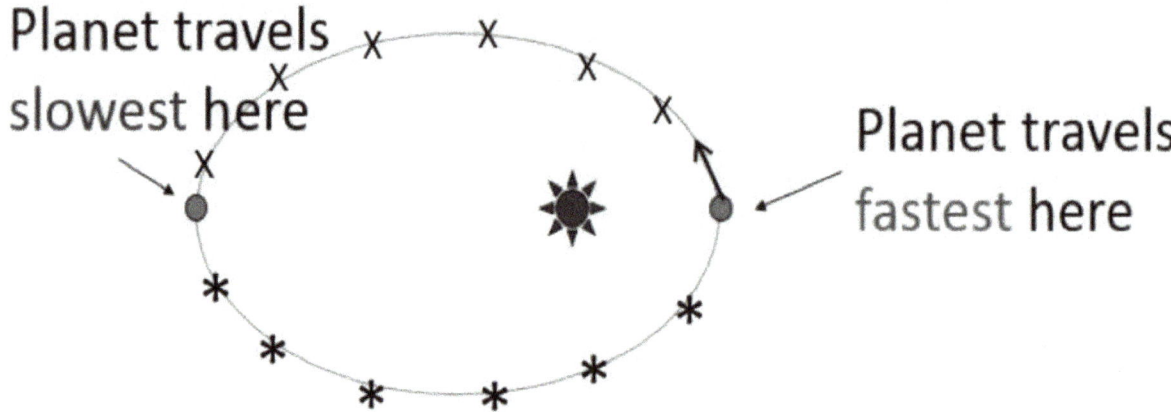

Figure 5.4: Variation in Planet Speed During Orbit

Lastly is Kepler's 3rd Law. This law tells us that a planet's orbital period squared is proportional to its semi-major axis cubed. In mathematical form, this is $p^2 = a^3$. p is the period in years; it is how long the planet takes to orbit the Sun once. The Earth's period, for example, is one year. a is the semi-major axis in AU; this is the planet's average distance from the Sun. The Earth's semi-major axis, for example, is 1 AU. Note that p^2 means the period times itself, and that a^3 means the semi-major axis times itself times itself again. In words, what the third law means is that planets that are closer to the Sun have shorter orbital periods (because they travel faster and have less distance to go) and planets that are further from the Sun have longer orbital periods (because they travel slower and have more distance to go). For example, in Figure 5.5(A) there are two planets orbiting the Sun, a big one and a small one. Since the big one is further away, it has a longer period. In Figure 5.5(B) both planets are the same distance from the Sun, so they have the same period. Note that the fact that one planet is bigger than the other is irrelevant. Kepler's 3rd Law tells us that the only factor that matters in determining the period is distance.

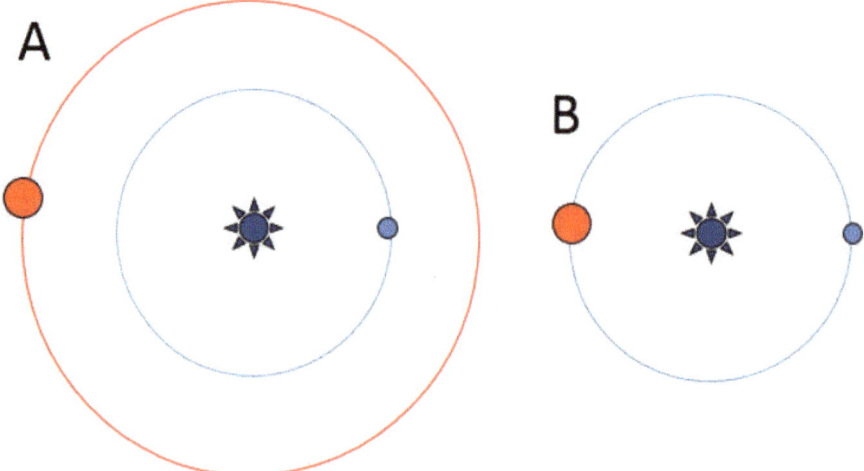

Figure 5.5: Kepler's 3rd Law Example

Figure 5.6 demonstrates Kepler's 3rd Law for the solar system. For each planet, the period and semi-major axis are listed. The fourth column shows the period squared, and the last column shows the semimajor axis cubed. Note that the last two columns are equal, in accordance with the third law.

	Period P (years)	Semi-major Axis a (AU)	P^2	a^3
Mercury	0.24	0.39	0.058	0.059
Venus	0.61	0.72	0.372	0.373
Earth	1	1	1.000	1.000
Mars	1.88	1.52	3.534	3.512
Jupiter	11.86	5.2	140.660	140.608
Saturn	29.46	9.54	867.892	868.251
Uranus	84.01	19.19	7057.680	7066.835
Neptune	164.79	30.06	27155.744	27162.324

Figure 5.6: Demonstration of Kepler's Third Law

Be sure not to confuse the 2nd and 3rd laws. Both deal with speed, but they refer to different things. The 2nd law tells us what a *particular* planet does as it orbits the Sun; when it is closer to the Sun it moves faster. The 3rd law compares the orbital periods for *all* of the planets in the solar system; planets that are further from the Sun travel slower and have longer orbital periods.

Questions You Should Be Able to Answer
1. What are Kepler's Three Laws?

Suggestions for Further Reading
1. http://astro.unl.edu/naap/pos/pos_background1.html (explanation of Kepler's Laws)
2. http://www.astro.washington.edu/users/smith/Astro150/Tutorials/Kepler/ (another explanation of Kepler's Laws)

Suggestions for Practice
3. http://astro.unl.edu/interactives/kepler/KeplerFirstLaw.html (1st Law Questions)
4. http://astro.unl.edu/interactives/kepler/KeplerSecondLaw.html (2nd Law Questions)
5. http://astro.unl.edu/interactives/kepler/KeplerThirdLaw.html (3rd Law Questions)
6. http://astro.unl.edu/classaction/questions/renaissance/ca_renaissance_kepler2nd.html (more 2nd Law Questions)

Chapter 6: Seasons

In this chapter, you will learn why we experience seasons.

The Earth is constantly moving. It rotates on its axis once a day, and it revolves around the Sun once a year.

The rotation of the Earth causes night and day. When you are on the side of the Earth that is facing the Sun, it is daytime. When you are on the side of the Earth facing away from the Sun, it is nighttime. Figure 6.1 shows the Earth's rotation as seen looking down on the north pole. The arrows represent the direction that the sunlight is coming from (the Sun is to the right in this diagram). Seen this way, the Earth rotates counterclockwise, and different times of day correspond to different positions in the rotation. For example, when you are directly facing the Sun it is noon, and when you are directly facing away from the Sun, it is midnight. When you are moving from the lit half of the Earth to the unlit half, it is sunset (about 6 pm), and when you are moving from the unlit half to the lit half, it is sunrise (about 6 am).

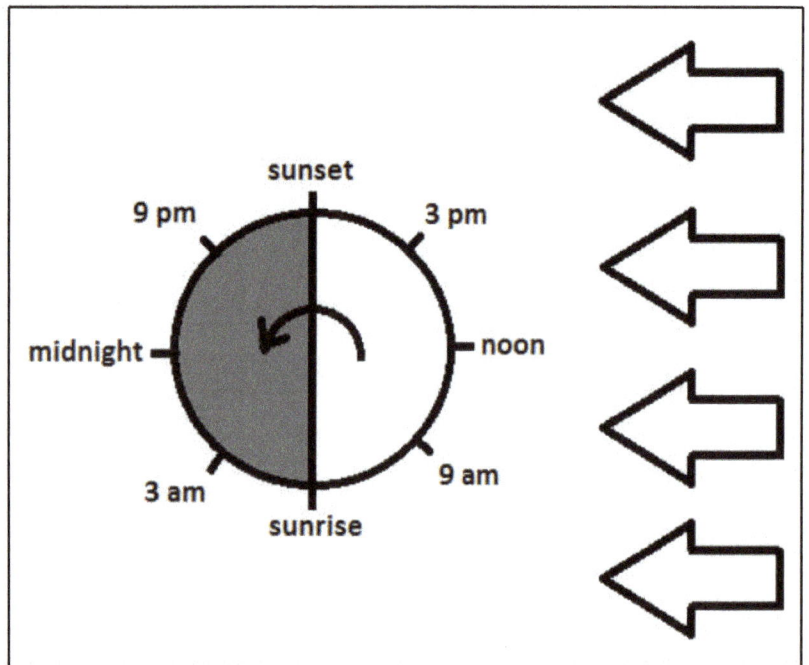

Figure 6.1: Earth's Rotation and Time of Day

The revolution of the Earth causes the seasons. There is a common misconception that this is a result of the Earth's elliptical orbit. A lot of people believe that we have seasons because we are closer to the Sun in the summer and further from the Sun in the winter. This is NOT true.

Remember the Earth's orbit to scale; it is almost a circle (Figure 6.2)! The ellipticity is not enough to cause significant temperature differences. While the distance between the Earth and the Sun does vary by 5 million km, this is a difference of only about 3% of the average Earth-Sun distance...essentially insignificant. In fact, if you look at the distance between the Earth and Sun (Figure 6.2), we are actually closer to the Sun in December, when we experience winter. Furthermore, when it is summer for us, it is actually winter for the southern hemisphere. The southern hemisphere experiences opposite seasons from us. This would not be the case if the Earth's elliptical orbit were what caused the seasons. If the elliptical orbit were what caused the seasons, both hemispheres would experience summer at the same time (when the Earth was closest to the Sun).

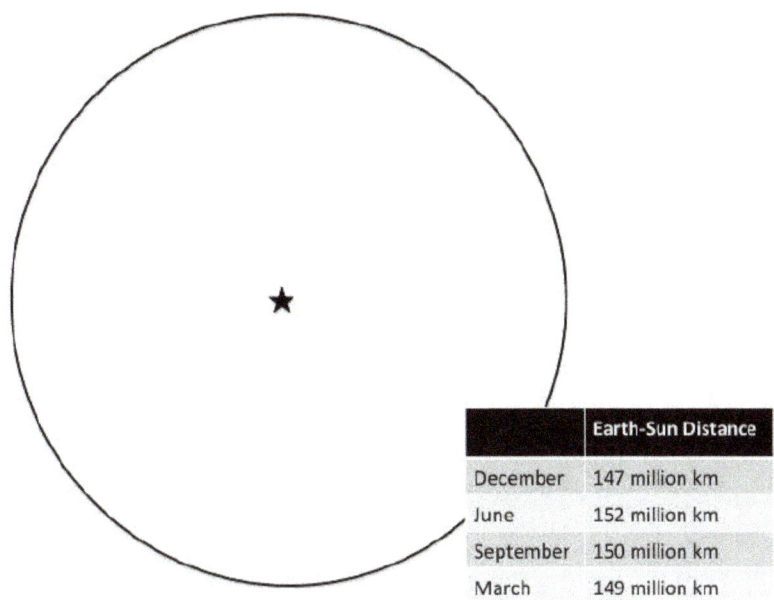

	Earth-Sun Distance
December	147 million km
June	152 million km
September	150 million km
March	149 million km

Figure 6.2: The Earth's Orbit to Scale

So what does cause the seasons? The answer is the Earth's tilt. The Earth's rotational axis is tilted 23.5 degrees from vertical. The direction and amount of tilt always remains the same (e.g. no matter what time of year, the axis is always tilted 23.5 degrees towards the same direction in space), but because the Earth revolves around the Sun, this means that at different times of year different parts of the Earth are tilted towards and away from the Sun (Figure 6.3). In June, the northern hemisphere is tilted towards the Sun and we experience summer while the southern hemisphere experiences winter. In December, the southern hemisphere is tilted towards the Sun and we experience winter while the southern hemisphere experiences summer. In spring and fall, neither hemisphere is tilted towards the Sun.

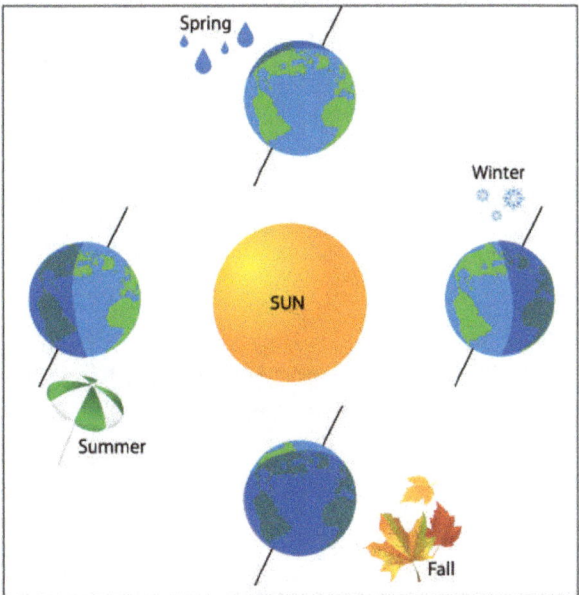

Figure 6.3: Seasons and the Earth's Tilt – Seasons Given for Northern Hemisphere (http://www.nasa.gov)

Why does the tilt matter? The tilt matters for two reasons. The first is the concentration of sunlight. When we are tilted towards the Sun, the Sun gets higher in the sky and the sunlight we receive is more concentrated (this is known as direct sunlight). When we are tilted away from the Sun, the Sun remains lower in the sky and the sunlight we receive is less concentrated (this is known as indirect sunlight).

This concept is illustrated in Figure 6.4. The line running across the Earth is the equator, and the dot halfway between the equator and the North Pole represents our location (NY). Note that in summer, the same beam of sunlight (represented by the flashlight) is spread out over less area than in the winter. Since it is spread out over a smaller area, it is more concentrated and it gets warmer.

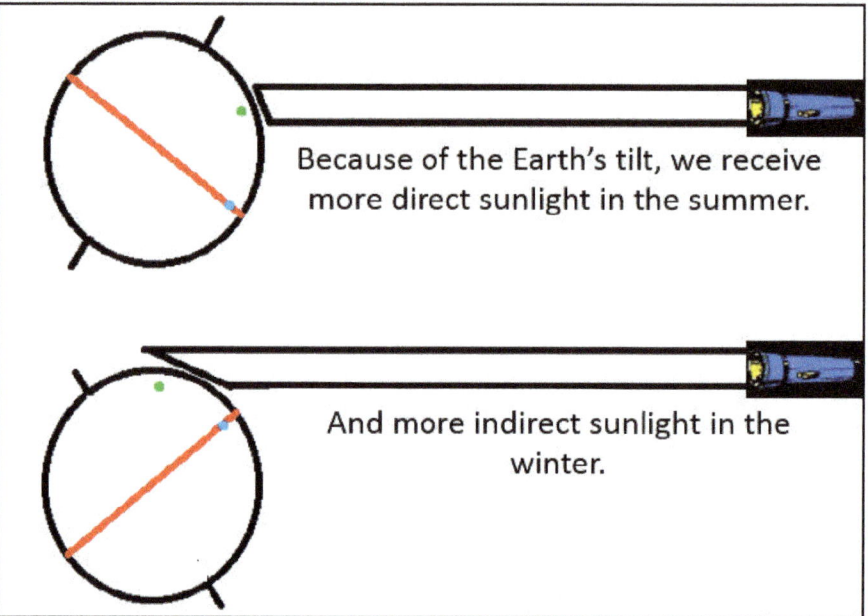

Because of the Earth's tilt, we receive more direct sunlight in the summer.

And more indirect sunlight in the winter.

Figure 6.4: Direct Versus Indirect Sunlight, NY

Now also look at a location on the equator (the dot on the equator; Figure 6.5). Note that for the location on the equator, there is not much difference in sunlight concentration between different times of the year...the sunlight is always concentrated and direct. This is why locations on the equator do not experience seasons; they are always warm. The further you go towards the poles, the more drastic the difference in sunlight concentration is between winter and summer, and the more drastic the seasonal change.

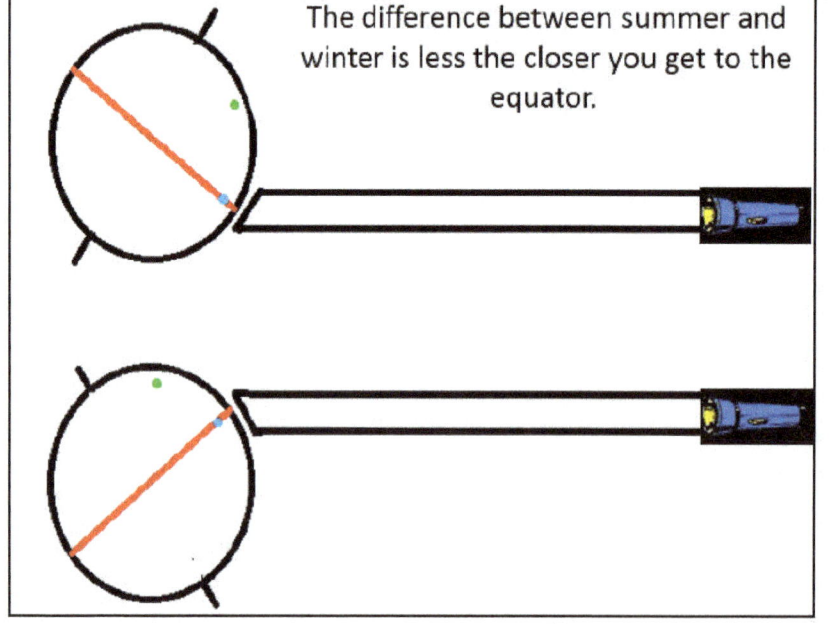

The difference between summer and winter is less the closer you get to the equator.

Figure 6.5: Direct Versus Indirect Sunlight, Equator

The second reason that the tilt matters is that it causes the number of daylight hours to vary (Figure 6.6). In June, the northern hemisphere gets proportionally more daylight hours (because it is tilted towards the Sun) and the southern hemisphere gets proportionally less (because it is tilted away from the Sun). The situation is reversed in December. So, for example, we (in NY, at 40 degrees north latitude) get about 15 hours of daylight in the summer, and 9 hours in the winter. In the spring and fall, everywhere on Earth gets the same amount of daylight (12 hours). Note that for the equator, the time of year once again does not matter; the equator always gets 12 hours of daylight. As you go towards the poles, the difference between summer and winter hours gets bigger. In fact, if you lived at the north or south poles, you would experience 6 months of daylight followed by 6 months of darkness! This is an additional reason why seasonal changes increase towards the poles and the equator does not experience seasons.

To sum up, when we are tilted towards the Sun we get more concentrated (direct) sunlight and more hours of sunlight. This makes it warmer and it is summer. When we are tilted away from the Sun we get less concentrated (indirect) sunlight and less hours of sunlight. This makes it colder and it is winter. Locations on the equator do not experience seasons because they always receive 12 hours of concentrated (direct) sunlight; thus they are always warm. Seasonal changes increase towards the poles.

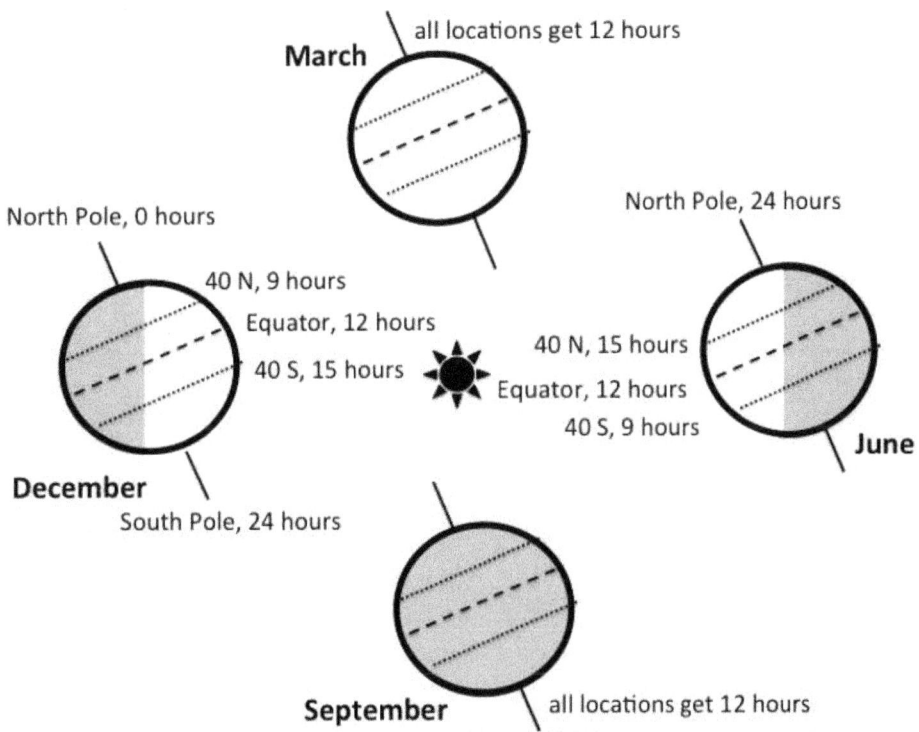

Figure 6.6: Variation in Number of Daylight Hours Throughout a Year

Questions You Should Be Able to Answer
1. Define rotation and revolution. How long does it take the Earth to complete a cycle of each?
2. What causes night and day?
3. What causes the seasons?

Suggestions for Further Reading
1. http://www.khanacademy.org/science/cosmology-and-astronomy/earth-history-topic/earth-title-topic/v/seasons-aren-t-dictated-by-closeness-to-sun (Khan Academy explanation of the seasons, part 1)

2. http://www.khanacademy.org/science/cosmology-and-astronomy/earth-history-topic/earth-title-topic/v/how-earth-s-tilt-causes-seasons (Khan Academy explanation of the seasons, part 2)
3. http://iqa.evergreenps.org/science/phy_science/tutorials/astrotutorials/Seasons/GoSeasons.html (seasons tutorial, plus practice questions)

Suggestions for Practice

4. http://astro.unl.edu/interactives/seasons/Season1_Distance.html (seasonal distance question)
5. http://astro.unl.edu/interactives/seasons/Season2_Flashlight.html (seasons direct/indirect light questions)
6. http://astro.unl.edu/interactives/seasons/Season3_illuminatedAreas.html (seasons illuminated areas questions)
7. http://astro.unl.edu/interactives/seasons/Season4_Time.html (seasons daylight time questions)
8. http://astro.unl.edu/interactives/seasons/Season5_Tilt.html (seasons tilt questions)

Chapter 7: Lunar Phases and Eclipses

In this chapter, you will learn why we see different phases of the moon and why we see eclipses.

You are probably aware that the moon goes through different shapes, or phases, over the course of a month. The moon starts completely dark. As the days progress, more and more of the moon becomes visible on the right-hand side until the moon is completely visible. Then the moon shrinks again, with the darkness coming in gradually from the left-hand side. Different moon phases have different names (Figure 7.1). If the moon is dark, this is known as a new moon. If the moon is completely lit, this is known as a full moon. When the moon is half lit, this is a quarter moon. If the half lit side is on the right, it is a 1st quarter, and if the half lit side is on the left, it is a third quarter. When the moon is between quarter and new, this is a crescent moon. If the lit up side is on the right, it is a waxing crescent moon. The term waxing means that the moon is getting bigger as time goes by. If the lit up side is on the left, it is a waning crescent moon. The term waning means that the moon is getting smaller as time goes by. When the moon is between quarter and full, this is a gibbous moon. If the lit up side is on the right, it is a waxing gibbous moon, and if the lit up side is on the left, it is a waning gibbous moon. (Note that things are slightly different in the southern hemisphere – in the southern hemisphere waxing and waning would be reversed, with waxing being when the moon was lit up on the left and waning being when the moon was lit up on the right!)

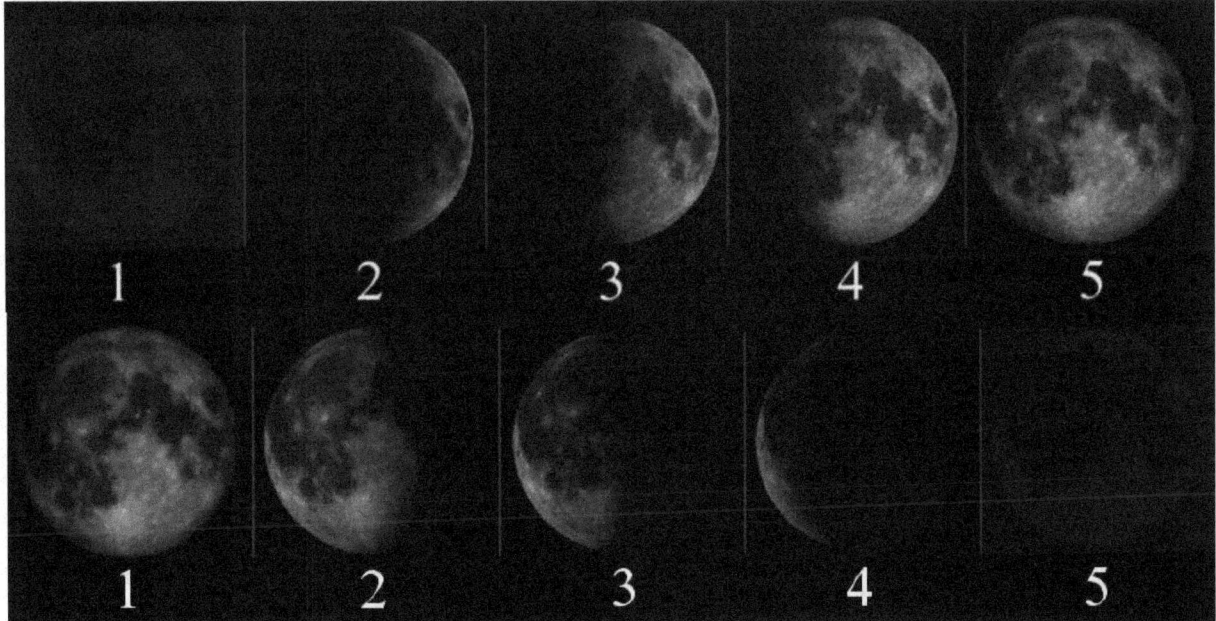

Figure 7.1: Moon Phases; Top Row → 1=new, 2=waxing crescent, 3=1st quarter, 4=waxing gibbous, 5=full
Bottom Row → 1=full, 2=waning gibbous, 3=3rd quarter, 4=waning crescent, 5=new
(https://commons.wikimedia.org/wiki/File:Waning_Moon_slides.jpg and
https://commons.wikimedia.org/wiki/File:Waxing_Moon_slides.jpg)

Figure 7.2 illustrates why we have moon phases. This shows the Earth-moon system looking down on the Earth's north pole. The moon's orbit around the Earth is shown (the inner circle represents the view that an alien would see looking down on the Earth's north pole, and the outer circle represents how someone on the Earth would see the moon). Note that the moon orbits counterclockwise around the Earth, taking one month (or 4 weeks) to go around once. In this diagram, the Sun's light is coming from the right. Notice that the half of the moon facing the Sun is always lit up. However, we on Earth see the half of the moon that is facing us, which is not always the lit up half! The moon's position in its orbit around the Earth determines how much of the lit up half we see, and hence the phase. For example, when the moon is between the Earth and the Sun, the unlit side of the moon is facing the Earth. This corresponds to the new moon phase. Half a week later, a little sliver of lit up moon is visible on the right. This corresponds to the waxing crescent phase. As time continues, a greater and greater proportion of the lit up side is visible. When the moon reaches the opposite side of the Earth from the Sun (2 weeks after the new moon phase), the entire lit up side is facing the Earth, and this is the full moon phase. The next two weeks correspond to the moon getting smaller, going back to the new moon phase.

Figure 7.2: Why We Have Moon Phases (http://www.nasa.gov)

A lot of people assume that the moon is only visible during the night. However, if you think about it, you can probably think of a time when you saw the moon during the day! This is because different phases rise and set at different phases. However, a given phase will always rise and set at the same time. For example, the full moon will always rise at 6 pm and set at 6 am, and the waning crescent moon will always rise at 3 am and set at 3 pm.

Figure 7.3 shows an example to see how this works. In 7.3A, we have the Earth (middle), the moon (left), and the Sun coming from the right. Again, we are looking down on the Earth's north pole. The moon's position corresponds to a full moon phase, because the lit side of the moon is facing the Earth. Since the Earth rotates counterclockwise once every 24 hours, a viewer on the Earth will see the Sun at different positions in the sky. This corresponds to different times of day, which have been labeled.

In 7.3B, we see a viewer for whom it is noon. Note that I've blocked out the left hand half of the picture. That's because our viewer's horizon runs vertically down the middle of the page; the black is blocking out everything that is below the horizon and hence can't be seen at that time. Everything not blocked out is above the horizon and hence can be seen. For instance, here the Sun is directly overhead our viewer's head, at its highest point in the sky. Since our viewer is facing south, their left hand side is the eastern horizon and their right hand side is the western horizon. Note that the moon is below the horizon. So while the moon is still there, and is still full, it can't be seen at noon because it hasn't risen yet.

Let's move 3 hours ahead (7.3C). It is now 3 pm for our viewer. Notice that our viewer's view of the sky has changed a bit. A part of the sky which was below the horizon before is now visible. A part of the sky which was above the horizon before is now below and is not visible. The Sun is lower in our viewer's sky. However, the moon is still not yet visible. 7.3D moves ahead another 3 hours. It is now 6 pm, and the Sun is setting in the west. Notice what the full moon is doing. It is just starting to rise on the eastern horizon! 3 hours later, at 9 pm, the full moon is still visible, but it is higher in the sky than it was before (7.3E). At midnight, the full moon is now directly overhead, and has reached its highest position in the sky (7.3F). At 3 am, the full moon is still visible, but is lower in the sky (7.3G). At 6 am, we see the full moon setting in the west, and the Sun rising in the east (7.3H).

Figure 7.4 shows what this would look like if you were standing on the Earth facing south. Since you are facing south, east would be to your left and west would be to your right. At 6 pm, you would see the full moon rise on the eastern horizon (note that objects in the sky always rise in the east and set in the west). As the night progressed, it would rise higher and higher in the sky until midnight, when it would reach its highest position over the southern horizon. After that time, the moon would progressively get lower in the sky, until it set at 6 am.

Now imagine another phase, like the waxing crescent phase, which would place the moon directly about the position corresponding to 3 pm on the Earth. It is clear that the waning crescent phase would be visible at 3 pm, even though the full moon would not be visible. Thus the position of the moon, and hence the phase, determines the rising and setting time!

35

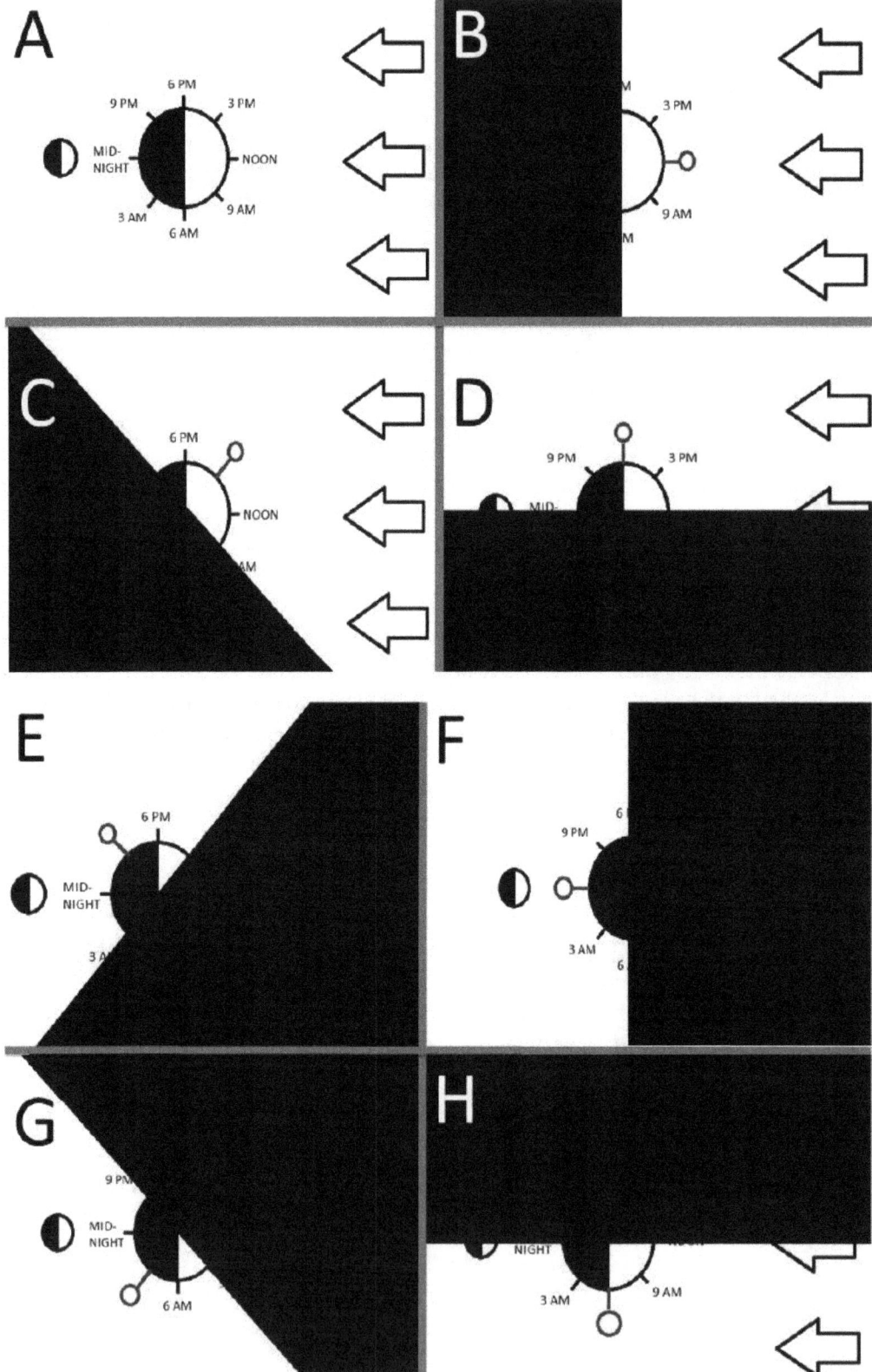

Figure 7.3: Full Moon Location Throughout Day

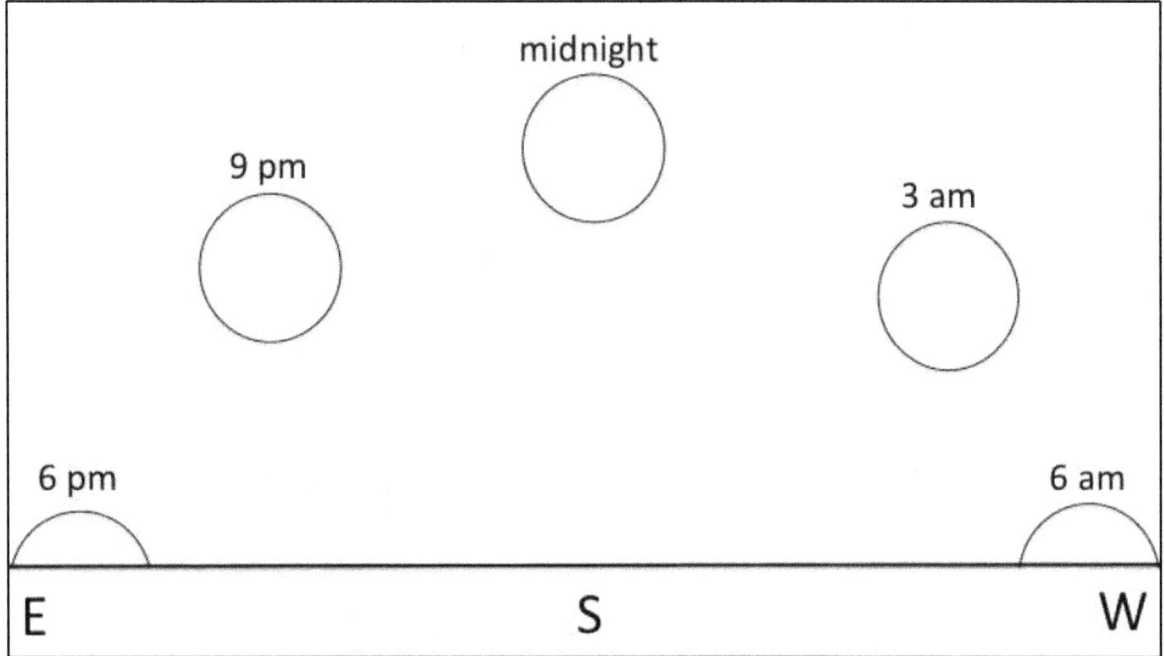

Figure 7.4: Full Moon's Position in Sky at Different Times

There is a simpler way to determine rising and setting times. You can make a diagram as shown in Figure 7.5, showing the Earth and the times on the Earth, and the different phases of the moon. Note that each phase is directly above a certain time on the Earth. This is the time that that phase appears highest in the sky. For example, the 3rd quarter phase is highest in the sky at 6 am and the waxing crescent phase is highest in the sky at 3 pm. Since the moon is above the horizon for 12 hours (half the day) you can then get the rising time by going backward (subtracting) 6 hours and the setting time by going forward (adding) 6 hours. Let's use the waxing gibbous phase an example. The waxing gibbous phase is highest in the sky at 9 pm. Thus is rises at 3 pm (6 hours earlier) and sets at 3 am (6 hours later).

When you are using Figure 7.5 to determine rising and setting times, remember that the time of day does *not* determine the phase. For example, just because it is 9 pm does not mean that it is a waxing gibbous phase. The phase of the moon does not change over a day; it takes about 3.5 days to go from one phase to another! Thus at 9 pm, the moon could be in any phase. The phase is determined by where the moon is in its orbit around the Earth; the time of day just determines whether or not you can see the moon and where in the sky you see it.

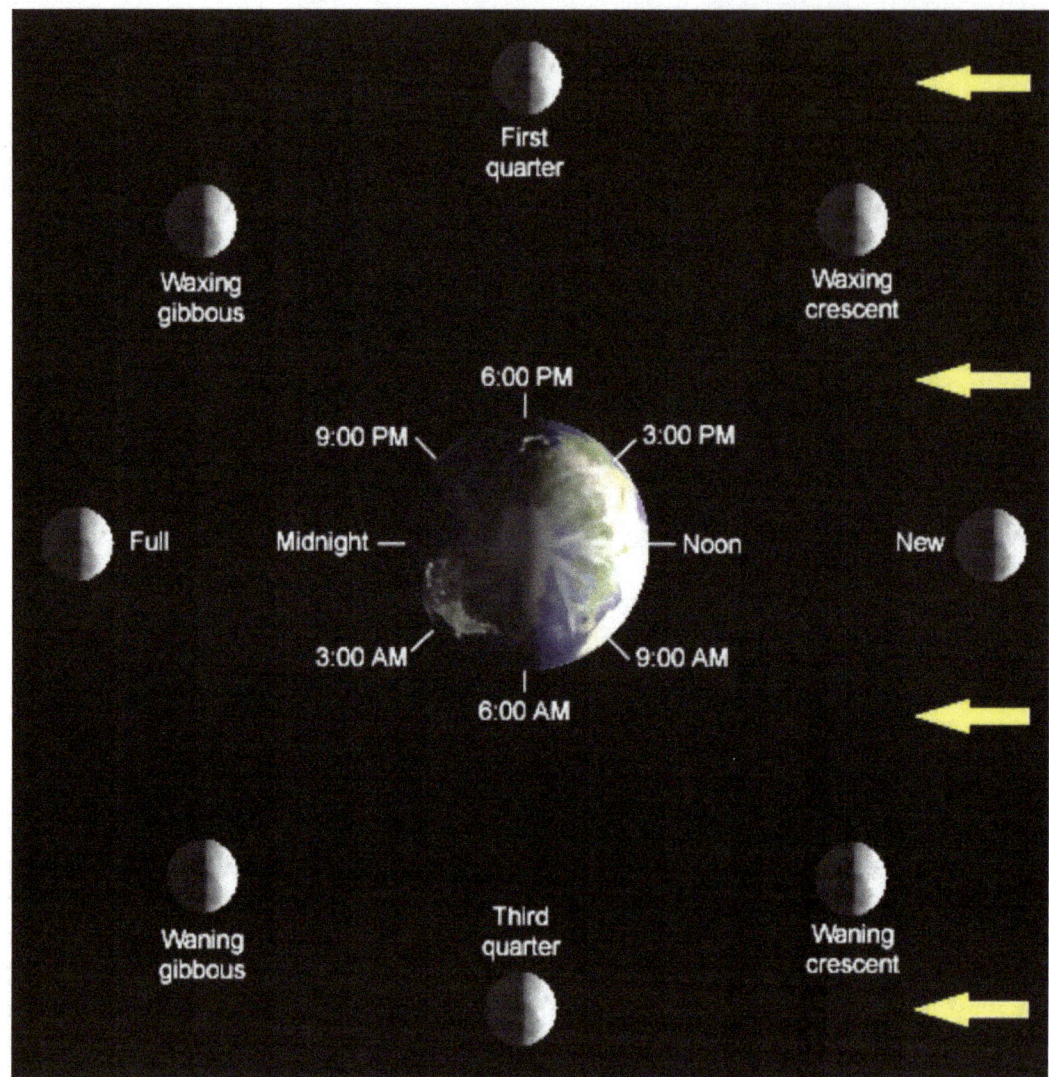

Figure 7.5: Moon Phases and Time of Day (https://commons.wikimedia.org/wiki/File:Lunar-Phase-Diagram.png)

Before we finish with lunar phases, let's take a moment to talk about eclipses. There are two types of eclipses: lunar eclipses and solar eclipses. A lunar eclipse occurs when the moon passes through the Earth's shadow (Figure 7.6). Since the Earth's shadow falls behind the Earth, a lunar eclipse will only occur when the moon is full. What you will see during a lunar eclipse is the moon slowly getting darker as the Earth's shadow falls across it. When the moon is entirely within the Earth's shadow, it will actually appear red. This is from light being bent through the Earth's atmosphere and reflected off of the moon. It looks red because red light is bent the most. Note that if there is a lunar eclipse, every location on the Earth that is having nighttime when the eclipse occurs will see it.

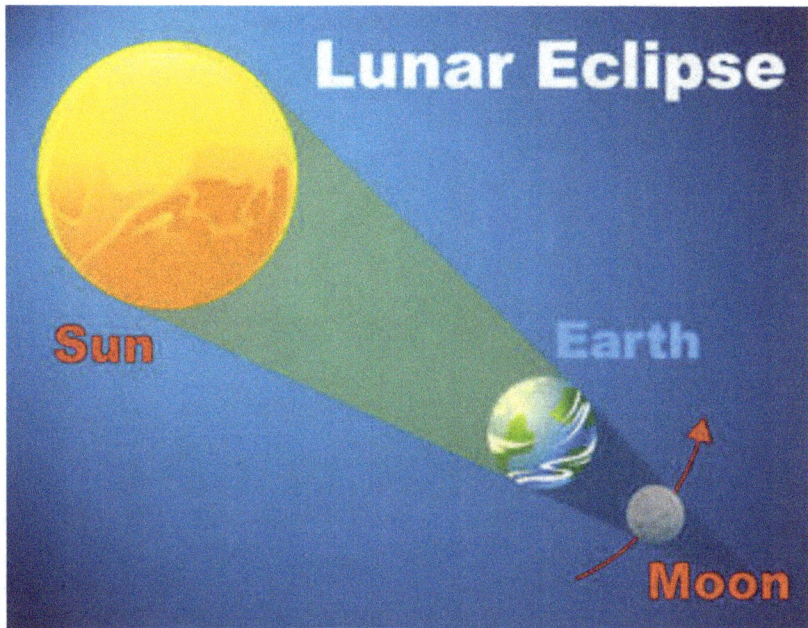

Figure 7.5: Configuration During A Lunar Eclipse (http://www.nasa.gov)

A solar eclipse occurs when the new moon passes between the Earth and the Sun and blocks the light from the Sun, casting a shadow on the Earth (Figure 7.6). Since the shadow only falls on a small portion of the Earth, you need to be within a narrow swath about 100 miles wide to see the eclipse. An interesting coincidence is that the Sun and the moon appear the same size in the Earth's sky. Remember the Sun is *actually* about 400 times bigger than the moon – but because it is so much further away, it appears the same size! This means that during a solar eclipse, the moon will exactly cover the Sun, making it completely dark (like nighttime) and leaving just the glow from the Sun's outermost layers visible (Figure 7.7).

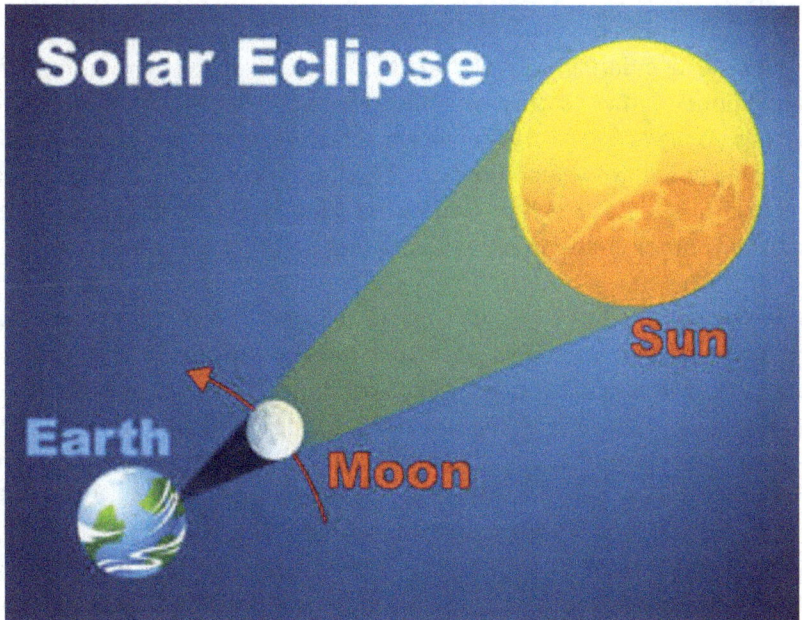

Figure 7.5: Configuration During A Solar Eclipse (http://www.nasa.gov)

Figure 7.6: Sun/Moon During a Solar Eclipse, 2017 Eclipse (http://www.nasa.gov)

This is not true for every planet! For example Figure 7.7 shows views from a solar eclipse as viewed from Mars; the Martian moon Phobos is eclipsing the Sun, but it is not as big as the Sun in the Martian sky. These photos were taken by the rover Curiosity on the surface of Mars.

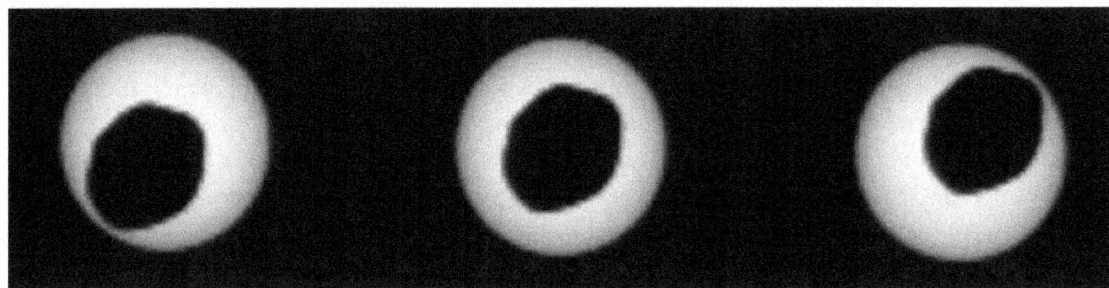

Figure 7.7: Solar Eclipse from the Surface of Mars (http://www.nasa.gov)

You might now be wondering why we don't have eclipses every month. This is because the Moon's orbit is slightly tilted relative to the plane between the Earth and the Sun (Figure 7.8). So usually the Earth's shadow falls above or below the full moon and the new moon's shadow falls above or below the Earth. There are only two times a year when the shadows are potentially aligned. If the moon is full during one of these two times, there will be a lunar eclipse. If the moon is new, there will be a solar eclipse. If the moon is any other phase, there will be either no eclipse or a partial eclipse.

Figure 7.9 shows recent upcoming total solar eclipses. In this time span, two fall across the continental US. The first occurred on August 21, 2017. Because it was the first total solar eclipse in the continental US since 1979, and because it covered an area ranging from coast to coast, it was nicknamed "The Great American Eclipse." Your professor was lucky enough to travel to Grand Teton National Park to view the eclipse from the region of totality (Figure 7.10), and can attest to the fact that it is one of the most amazing things you can ever see! Which brings us to your next chance to see a total solar eclipse in the US – which is on April 8, 2024. The region of totality for this eclipse will pass through portions of Texas, Oklahoma, Arkansas, Missouri, Illinois, Kentucky, Indiana, Michigan, Ohio, Pennsylvania, Vermont, New Hampshire, Maine, and even (a very very tiny portion near Buffalo) New York. Your next chance in the US after that will not be until 2044!

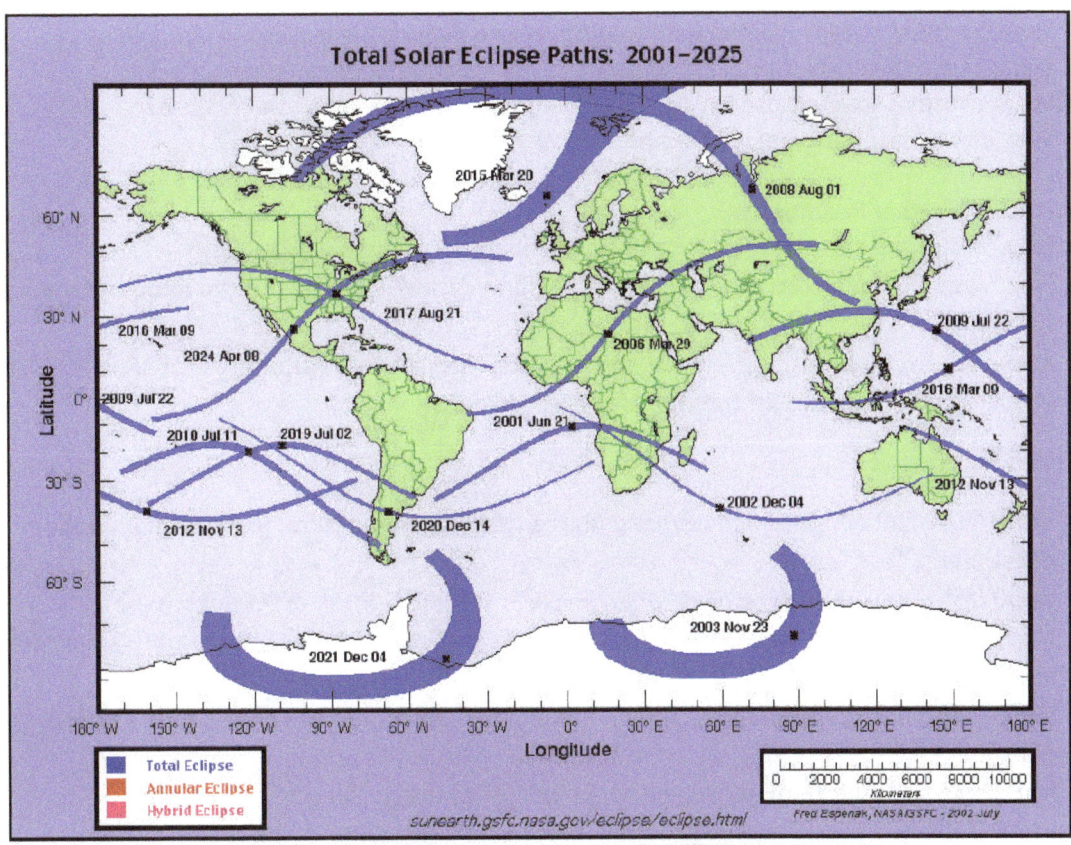

Figure 7.9: Upcoming Solar Eclipses (http://www.nasa.gov)

Figure 7.10: Your Professor Views the "Great American Eclipse" of 2017

Questions You Should Be Able to Answer

1. What causes lunar eclipses?
2. Given the position of the moon, Earth, and Sun, indicate the phase of the moon.

3. For each lunar phase, give the time that the phase rises, is highest in the sky, and sets.
4. What causes a lunar eclipse? What phase is the moon in during a lunar eclipse?
5. What causes a solar eclipse? What phase is the moon in during a solar eclipse?
6. Why don't lunar and solar eclipses happen every month?

Suggestions for Further Reading
1. http://astro.unl.edu/naap/lps/lunarPage1.html (moon phases explanation)
2. http://astro.unl.edu/classaction/animations/lunarcycles/lunarapplet.html (moon phases simulator)
3. http://iqa.evergreenps.org/science/phy_science/tutorials/astrotutorials/MoonPhases/GoMoon Phases.html (moon phases tutorial, plus practice questions)

Suggestions for Practice

4. http://astro.unl.edu/interactives/lunar/PercentOfIllumination.html (percent illumination questions)
5. http://astro.unl.edu/interactives/lunar/PhaseOrder.html (phase order questions)
6. http://astro.unl.edu/interactives/lunar/PhaseOrder_GeometricView.html (more phase order questions)
7. http://astro.unl.edu/interactives/lunar/PhaseOrder_MultipleViews.html (more phase order questions)
8. http://astro.unl.edu/interactives/lunar/EarthTimes.html (Earth times questions)
9. http://astro.unl.edu/interactives/lunar/EarthTimes3D.html (Earth times 3D questions)
10. http://astro.unl.edu/interactives/lunar/MeridianTimes.html (meridian times questions)
11. http://astro.unl.edu/interactives/lunar/RisingSettingTimes.html (rising and setting times questions)
12. http://astro.unl.edu/interactives/lunar/Flat_Horizon_View_SamePhase.html (horizon view questions)
13. http://astro.unl.edu/interactives/lunar/Flat_Horizon_View_sameStartTime.html (more horizon view questions)
14. http://astro.unl.edu/interactives/lunar/Flat_Horizon_View_AllDifferences.html (more horizon view questions)
15. http://astro.unl.edu/interactives/lunar/PhaseOrder_3DSunFixed.html (3D questions)
16. http://astro.unl.edu/interactives/lunar/PhaseOrder_3DSunVariable.html (more 3D questions)
17. http://astro.unl.edu/interactives/lunar/Lunar1_ST.html (sorting question 1)
18. http://astro.unl.edu/interactives/lunar/Lunar2_ST.html (sorting question 2)
19. http://astro.unl.edu/interactives/lunar/Lunar3_ST.html (sorting question 3)
20. http://astro.unl.edu/classaction/questions/lunarcycles/ca_lunarcycles_idphase.html (phase ID picture questions)
21. http://astro.unl.edu/classaction/questions/lunarcycles/ca_lunarcycles_idlunarpos.html (phase ID position questions)
22. http://astro.unl.edu/classaction/questions/lunarcycles/ca_lunarcycles_horzdiagidphase.html (phase ID horizon diagram questions)
23. http://astro.unl.edu/classaction/questions/lunarcycles/ca_lunarcycles_horzdiagidtime.html (time ID questions)

Chapter 8: Origin of the Solar System

In this chapter, you will learn about the prevailing hypothesis for the formation of the solar system, the solar nebula hypothesis. You will also learn about a process called differentiation and about the different types of atmospheres planets can have.

Recall that the Earth is one of eight planets in our solar system. Our Sun is one of hundreds of millions of stars in our galaxy, and our galaxy is one of hundreds of millions of galaxies in the universe. Here we are going to assume we have a galaxy and see how we go from a cloud of gas and dust in a galaxy into the solar system.

We are first going to see what observations scientists used when constructing a hypothesis for solar system formation. Any good model of solar system formation will need to explain these observations. I will then present the hypothesis, and in class we will discuss how these observations fit in with the hypothesis.

Our first observation is that all of the planets orbit the Sun in the same direction, in the same plane, and in nearly circular orbits. You can envision the solar system as a record with the Sun at the center and the planets moving counterclockwise around the center in the record grooves.

Not only do the planets orbit the Sun counterclockwise, but they also spin counterclockwise on their axes. The two exceptions are Uranus, which is tilted about 90 degrees on its axis, and Venus, which spins very slowly clockwise. Venus' spin is so slow that it takes 242 Earth days for it to rotate once!

Our third observation is that the planets can be divided into two groups, with the small rocky bodies (the Terrestrial planets) located close to the Sun and the large gaseous bodies (the Jovian planets) located far from the Sun.

Our final observation is that we have leftover debris in the solar system in the form of asteroids, KBOs, and comets.

These observations have led scientists to the prevailing hypothesis of planet formation, the solar nebular hypothesis. This hypothesis states that our solar system formed from the collapse of a large, spinning, spherical cloud of gas and dust.

This process is illustrated in Figure 8.1 (at end of chapter). The solar system begins as a large, spherical, spinning cloud of gas and dust. About 5 billion years ago, this cloud begins to collapse. Scientists believe that the shockwave from a nearby supernova explosion may have triggered the collapse. Just as an ice skater spins faster as she brings her arms closer to her body, the cloud spins faster as it collapses. As collapse continues, the cloud flattens out into a disk shape. Gravity causes most of the mass to concentrate in the center of the disk. As collapse proceeds, temperatures in the disk increase. Eventually temperatures in the center are hot enough that nuclear fusion begins and the Sun ignites. In the outer regions of the disk, small regions of matter start to collect and clump together. Nearer to the Sun, where temperatures are above boiling, these clumps include rock and metal. Further from the Sun, where temperatures are below freezing, these clumps also include different ices, like water ice, methane ice, and ammonia ice. These small regions of collapsing matter grow into moon-sized protoplanets. The protoplanets grow into planets through collisions until most of the debris has been swept up. Because the outer regions of the solar system contain ice, the planets here grow faster and get so big that they are able to pull in gases (like hydrogen and helium) from the disk. Wind from the Sun eventually blows the remaining gas away from the newborn solar system. The timeline for solar system formation is roughly 100 million years.

Once formed, the early planets were very hot and mostly molten liquid. This is because of the heat generated by impacts during late stage planet formation as well as heat generated by short-lived radioactive isotopes that are no longer around.

Because the early planets were hot, this allowed a process called differentiation to occur. Differentiation is when a planet goes from being homogenous and well-mixed to being layered by density. To understand differentiation, imagine that you put some sand, water, and oil in a soda bottle, put a cap on the bottle, shake it up, and then set it down. The sand will sink to the bottom because it is heavier and the oil will rise to the top because it is lighter. The same thing happens with planets. The heavy material (the metal) will sink to the center, forming the core. The lighter material (the low density rock) will rise to the surface, forming the crust. The medium material (the medium density rock) will remain in the middle, forming the mantle.

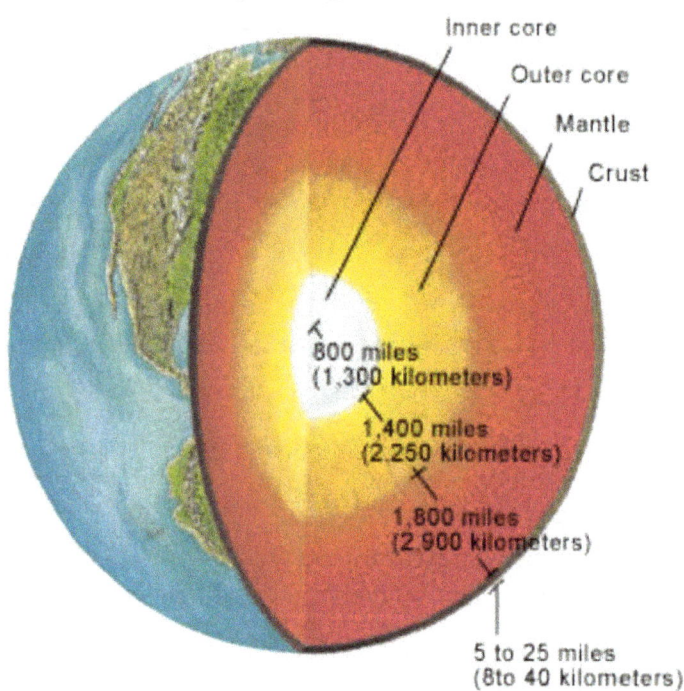

Figure 8.2: Layers of the Earth
(https://commons.wikimedia.org/wiki/File:Earth_layers_NA
SA.png)

Figure 8.2 illustrates this for the Earth. The surface of the Earth is a layer of low-density rock named the crust. Note that this layer is very thin (only about 5 to 25 miles thick). To give you a sense of scale, the thickness of the crust relative to the whole Earth is about the same ratio as the thickness of an apple's skin relative to the whole apple! Underneath the crust is the mantle, which is made of high-density rock and is about 1,800 miles thick. At the center is the metallic core, which is the inner 2,200 miles. The core consists of two parts – the outer core is liquid and the inner core is solid.

There are three different types of atmospheres that a planet can have. A primary atmosphere is one that is collected during planet formation and is composed primarily of hydrogen and helium. The Jovian planets have primary atmospheres; the Terrestrial planets don't because they didn't become big enough fast enough during planet formation to accumulate significant amounts of hydrogen and helium. An example of a primary atmosphere is Jupiter.

A secondary atmosphere is created from outgassing during differentiation. If you think back to the bottle of sand, water, and oil, and make the water carbonated, you will see an analogy to how this happens. If you shake up the mixture, take the cap off, and set the bottle down, as the sand sinks and the oil rises, the carbon dioxide will seep out.   The same thing happened during planetary differentiation. Gases dissolved in the molten liquid seeped out, were held in by the planets' gravity, and became the secondary atmosphere. This gas was mostly carbon dioxide with some nitrogen. An example of a planet with a secondary atmosphere is Venus.

The Earth's early atmosphere looked a lot like Venus' atmosphere. However, the Earth's atmosphere underwent significant modification, or changes. Hence we call it a tertiary atmosphere. Much of the carbon dioxide in the Earth's atmosphere was removed as the Earth's oceans absorbed the carbon dioxide and stored it in rocks. Then oxygen was added as plants developed and emitted oxygen as part of photosynthesis. The result is that the Earth's atmosphere is composed of about 78% nitrogen and 21% oxygen, and is nearly 100 times thinner than Venus' atmosphere.

Questions You Should Be Able to Answer

1. What observations led to the solar nebula hypothesis?
2. Describe the Solar Nebula Hypothesis.
3. Why did some planets develop into terrestrial planets and some into Jovian planets?
4. Why were early planets hot?
5. What is differentiation?
6. What is the difference between a primary, secondary, and tertiary atmosphere? Where in the solar system can I find examples of each?

Suggestions for Further Reading

1. https://www.youtube.com/watch?v=Uhy1fucSRQI (Stephen Hawking explains the formation of the solar system in less than 5 minutes)
2. http://iqa.evergreenps.org/science/phy_science/tutorials/astrotutorials/SolSysForm/GoSolSysForm.html (solar system formation tutorial, plus practice questions)

How did our solar system come to be?

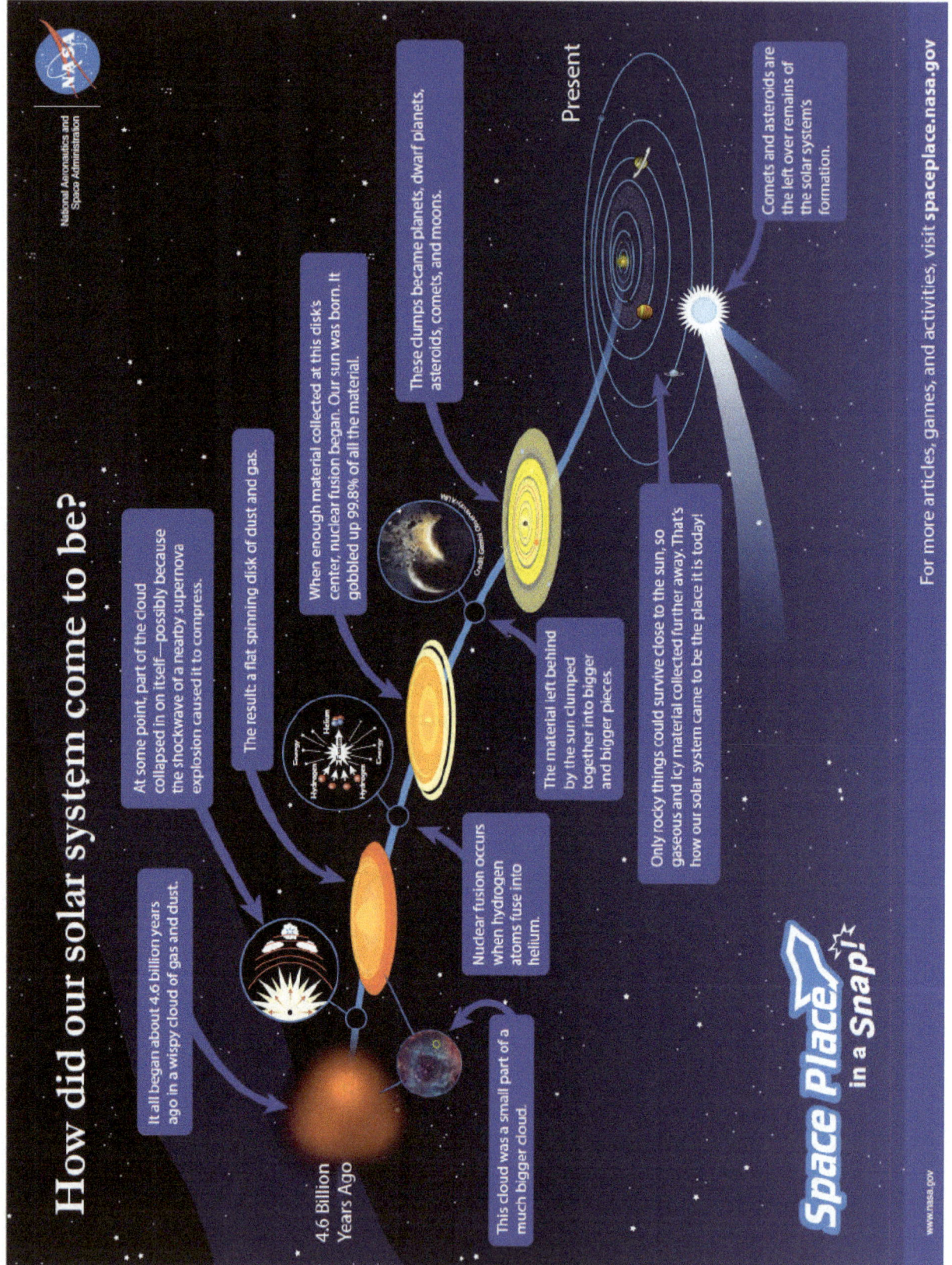

4.6 Billion Years Ago

It all began about 4.6 billion years ago in a wispy cloud of gas and dust.

This cloud was a small part of a much bigger cloud.

At some point, part of the cloud collapsed in on itself—possibly because the shockwave of a nearby supernova explosion caused it to compress.

The result: a flat spinning disk of dust and gas.

Nuclear fusion occurs when hydrogen atoms fuse into helium.

When enough material collected at this disk's center, nuclear fusion began. Our sun was born. It gobbled up 99.8% of all the material.

The material left behind by the sun clumped together into bigger and bigger pieces.

Only rocky things could survive close to the sun, so gaseous and icy material collected further away. That's how our solar system came to be the place it is today!

These clumps became planets, dwarf planets, asteroids, comets, and moons.

Present

Comets and asteroids are the left over remains of the solar system's formation.

National Aeronautics and Space Administration

Space Place in a Snap!

www.nasa.gov

For more articles, games, and activities, visit spaceplace.nasa.gov

Chapter 9: Shaping Planetary Surfaces

In this chapter, you will learn about the different processes that shape the surfaces of the Terrestrial planets, and will come to an understanding of why the surfaces of the Terrestrial planets are so varied.

Shown in Figure 9.1 is a topographic map of the Earth's surface. Dark grays are low elevations, light grays are high elevations, and medium grays are medium elevations. If you look at this map carefully, several things should strike out at you. There are higher elevation continents and lower elevation oceans. Ridges, or underwater mountain chains, run through the center of each of the ocean basins. Mountain chains appear to be located primarily along the edges of continents. We would like to be able to explain why these observations are so; e.g. what makes the surface of the Earth look that way it does?

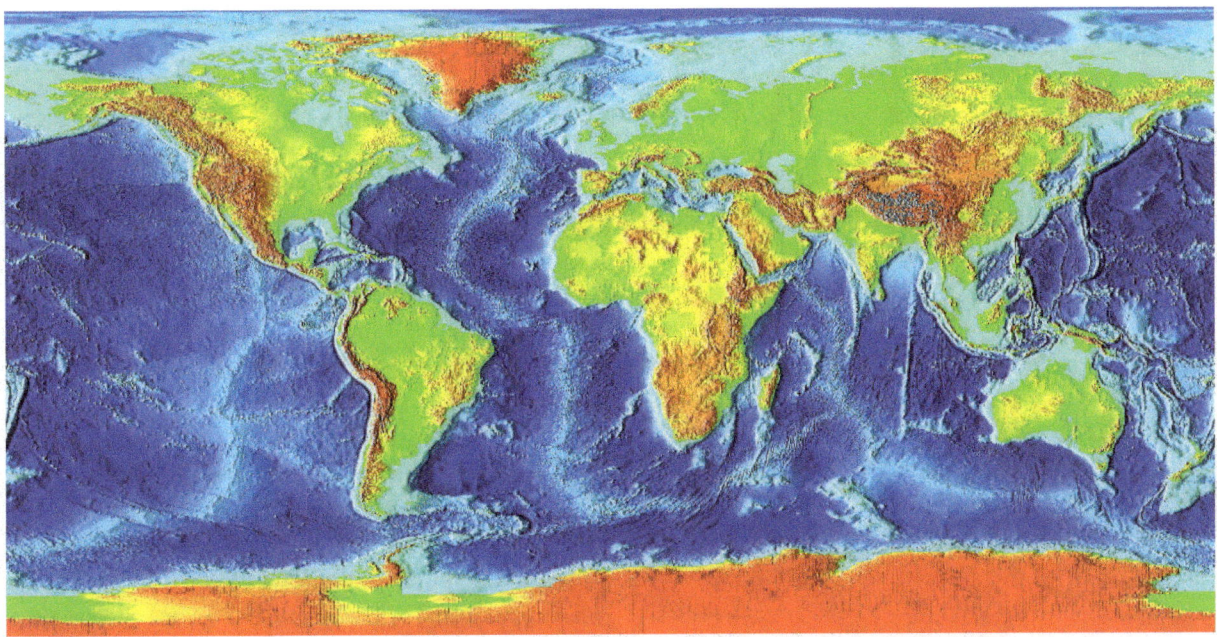

Figure 9.1: Topographic Map of the Earth (http://www.nasa.gov)

In contrast, Figure 9.2 shows a topographic map of Mars' surface. It looks very different from Earth's map. There are no continents, and the planet looks like its been almost divided in half, with higher elevations in the south and low elevations in the north. There are no ridges through the lower elevations. Mountains do not appear in chains, but appear scattered randomly. Circular depressions, called craters, are common. Thus we would like to explain not just what makes the surface of the Earth look the way it does, but also way makes the surface of Mars, and the other Terrestrial planets, look the way they do. Most importantly, we want to understand why the planets looks so different from eachother!

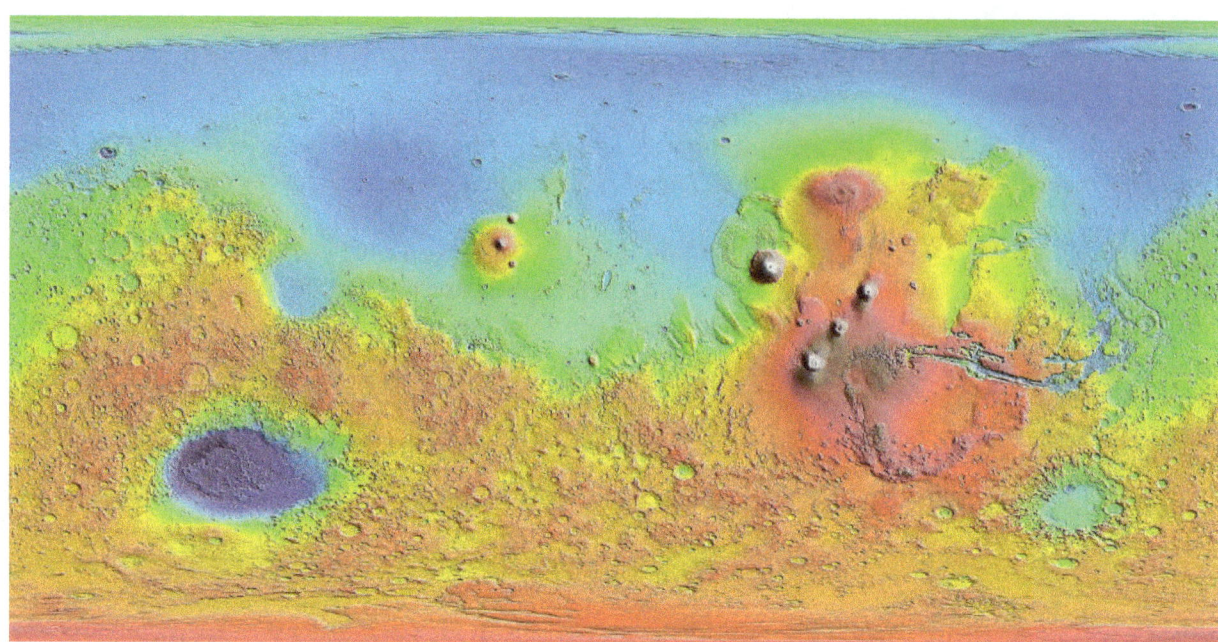

Figure 9.2: Topographic Map of Mars (http://www.nasa.gov, Mars Global Surveyor)

Five major processes are responsible for shaping the surfaces of most solid surfaces in the solar system. These processes are impact cratering, plate tectonics, volcanism, wind, and water. We will discuss each of these processes in turn.

Figure 9.3: Meteor Crater, Arizona (USGS)

Impact craters are created when rocks from space hit the surface of a planet. The collision creates an explosion, similar to a nuclear bomb detonating, and leaves behind a bowl-shaped scar in the ground. This bowl-shaped scar is called an impact crater. Figure 9.3 shows an example of an impact crater on the Earth. Note that there is no actual pieces of the rock from space that caused the crater left; the temperature and pressures during an impact event are so high that the rock form space gets vaporized. The map shown in Figure 9.4 gives you the locations of all known impact craters on the Earth's surface. You might notice that it looks as if there are more craters in North America and Europe, but this is probably an observational bias; there are more scientists in those areas looking for craters. However, it is clear that craters on the Earth are relatively rare.

This is not the case for all Terrestrial bodies. In fact, cratering is THE dominant process on many objects, for example Mercury and the Moon (Figure 9.5). When you look at these surfaces, all you see is crater upon crater upon crater!

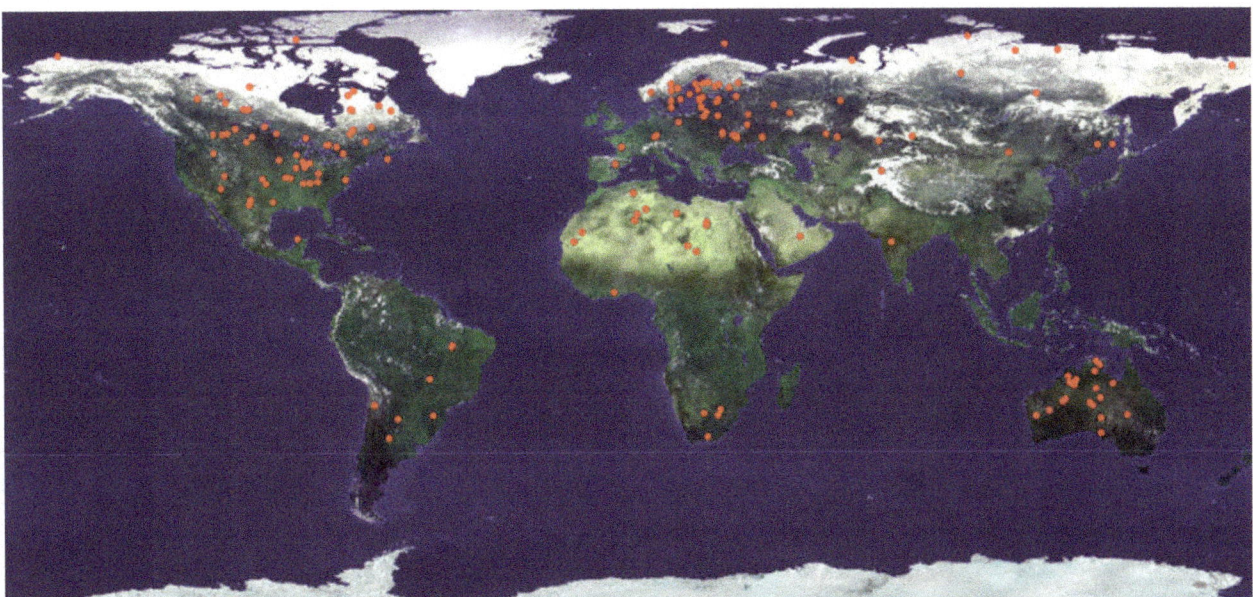

Figure 9.4: Locations of Impact Craters on the Earth (NASA Lunar and Planetary Institute)

Figure 9.5: The Cratered Surface of Mercury (http://www.nasa.gov, MESSENGER)

The process that is responsible for creating most of the features that we see on the Earth's surface is plate tectonics. The theory of plate tectonics says that the rigid outer surface layer of the Earth is cracked into plates, similar to a hardboiled egg where the shell is all cracked up. This layer is about 100 km thick; analogous to the thickeness of an apple's skin relative to the whole apple. The Earth's plates float around on a warm and weak underlying layer at a rate of about a few cm/yr (this is the same rate that your fingernails grow). As this happens, heat seeps out of the hot, partially molten interior, and plates interact at their boundaries. It is these interactions that create most of the physical features we see at the Earth's surface. The Earth is the only planet that we know of that has plate tectonics. However, recently scientists have discovered that Jupiter's moon Europa may have plate tectonics – with plates made of ice, not rock!

There are three ways that plates can interact at their boundaries, corresponding to three types of plate boundaries. At a divergent plate boundary, plates are moving away from each other. At a convergent plate boundary, plates are moving towards each other. At a transform plate boundary, plates are sliding past one another without colliding or splitting.

Each type of plate boundary is associated with different physical features (illustrated in Figure 9.6). Divergent boundaries typically occur in the middle of the ocean. As the ocean splits apart, magma rises up through the gap, creating a ridge and new seafloor. Thus if you remember back to the topographic map of the Earth, those underwater mountain chains in the middle of the oceans are actually divergent boundaries (often called "divergent ridges"). Underwater volcanism and earthquakes are also associated with divergent boundaries. At a convergent boundary, plate collide and generate mountains, volcanoes, deep cracks in the sea floor called trenches, and earthquakes. At transform plate boundaries, earthquakes are generated. This is why there are earthquakes in southern California. There, the North American plate slides past the Pacific plate along the San Andreas fault.

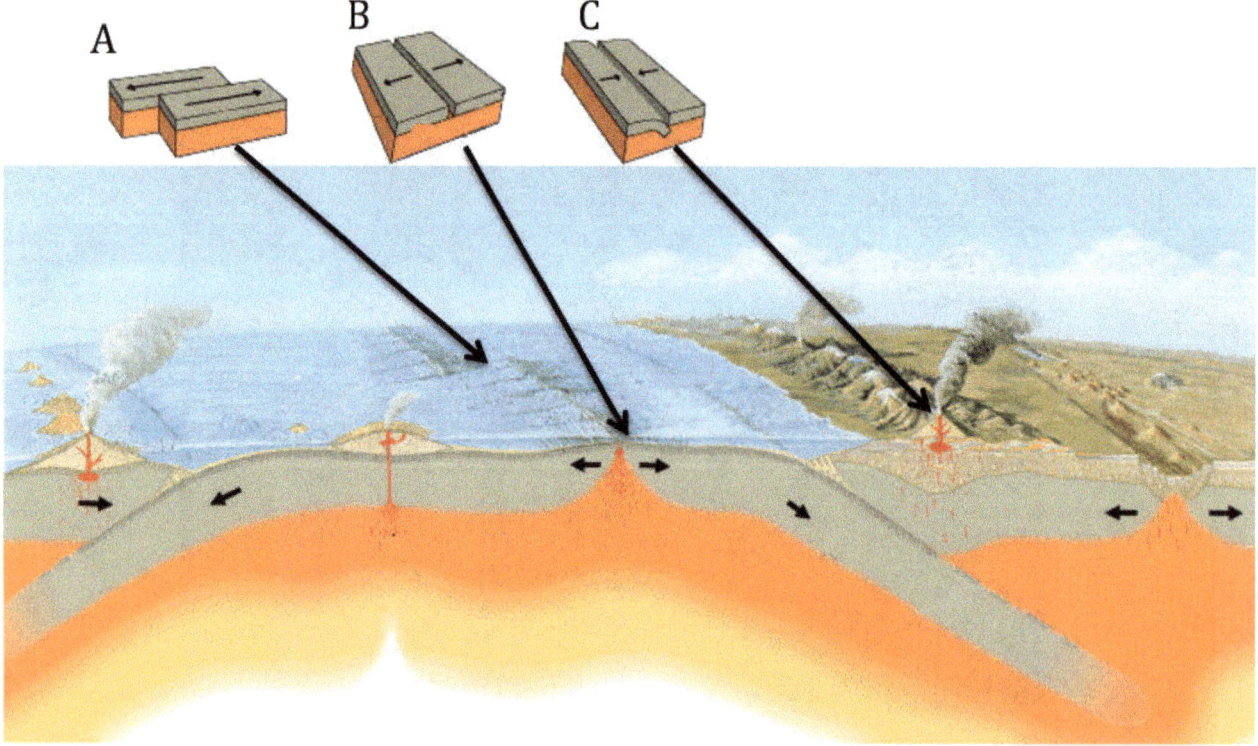

Figure 9.6: Three Types of Plate Boundaries, A is a transform boundary, B is a divergent boundary, and C is a convergent boundary (https://commons.wikimedia.org/wiki/File:Tectonic_plate_boundaries_clean.png)

Though plate tectonics can generate volcanoes, you don't need plate tectonics to have a volcano. You can also have non-plate tectonic related volcanism at what are termed hot spots. At a hot spot, there is a portion of the interior that is hotter than normal and acts like a flame underneath the surface, creating a volcano. An example of a hot spot on the Earth is Hawaii. An example of hot spot volcanism elsewhere in the solar system is Jupiter's moon Io. In Figure 9.7, we can see not one erupting hot spot volcano on Io, but two!! Hot spot volcanism is an alternative way for a planet to release heat when it doesn't have plate tectonics. Scientists don't yet know why some planets release their heat via plate tectonics and others solely via hot spot volcanism.

Figure 9.7: Erupting Volcanoes on Io (http://www.nasa.gov, New Horizons)

Wind can erode and deposit material, forming features like sand dunes (Figure 9.8) and wind streaks (Figure 9.9).

Water can also erode and deposit material. Figure 9.10 shows a stream bed on Mars. Note that it is covered with craters; this indicates that the stream bed is old...water has not flowed through the stream bed for a long time.

We can put what we've learned together and sum up what different features tell us about a planet. Volcanoes indicate that a planet has or had a hot molten interior. Divergent ridges indicate that a planet has or had a hot molten interior and plate tectonics. Sand dunes indicate that a planet has or had an atmosphere.

Figure 9.8: Sand Dunes on Mars (http://www.nasa.gov, Mars Reconnaissance Orbiter)

Stream beds indicate that a plate has or had water. Lastly, craters indicate that a feature is old. For example, if we see a volcano with a lot of craters on it, we know that that volcano has not erupted for a long time. Thus there once was a hot interior, but not any longer.

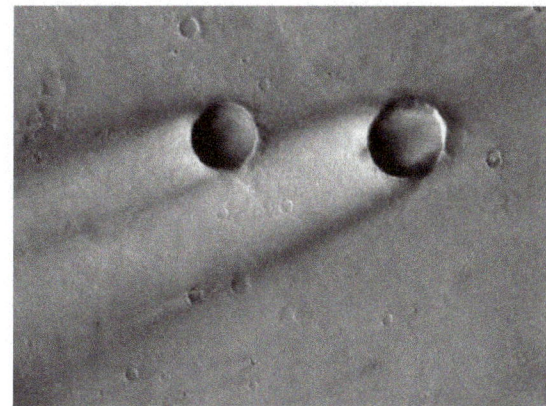

Figure 9.9: Dust Streaks on Mars (http://www.nasa.gov, Mars Odyssey)

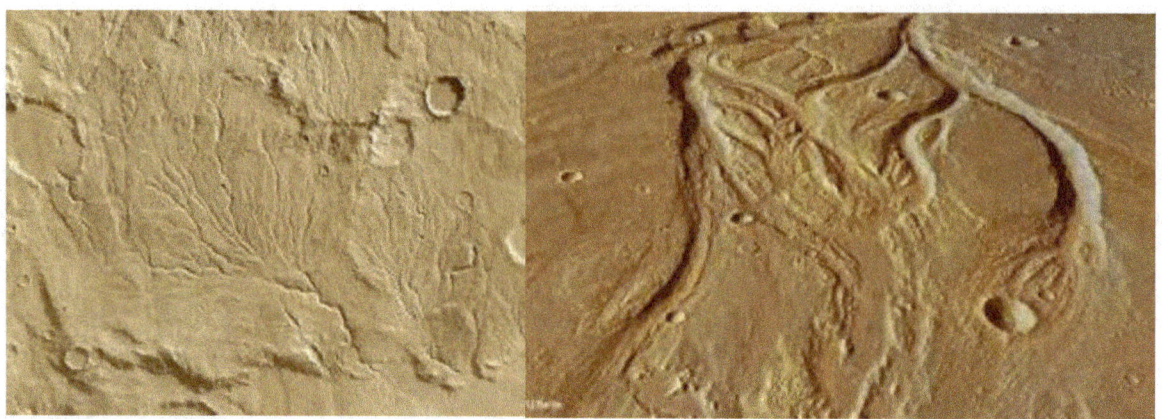

Figure 9.10: Stream Beds on Mars (http://www.nasa.gov; left - Viking Orbiter, right - Mars Express)

Questions You Should Be Able to Answer
1. What five major processes shape planetary surfaces?
2. What does the theory of plate tectonics say?
3. What is a hot spot? How do hot spots differ on a planet with plate tectonics and a planet without plate tectonics?
4. What do volcanoes, divergent ridges, sand dunes, stream beds, and craters tell us about a planet?
5. What do we mean when we say that a planet or moon is "geologically dead?"
6. How is size related to a planet's geologic activity?

Suggestions for Further Reading
1. http://www.lpi.usra.edu/education/explore/shaping_the_planets/background/ (shaping the planets reading)
2. http://iqa.evergreenps.org/science/phy_science/tutorials/astrotutorials/Shaping/GeoProcess.html (shaping planetary surfaces tutorial, with practice questions)

Chapter 10: Planetary Compositions

In this chapter, you will learn what planets are composed of. We will talk about what makes up the surfaces of the planets (rocks), and then we will talk about how scientists study the interiors of planets.

There are three types of rocks that can be found on the surface of a planet: igneous, sedimentary and metamorphic. Each rock type forms in a different way, and so tells us something different about the forces active on a planet (Figure 10.1).

	Source Material	Rock-Forming Process	Examples
Igneous	melting of rocks	crystallization (cooling and solidification of melted rock)	basalt, granite
Sedimentary	weathering and erosion of rocks into small bits (sediment)	deposition, burial, and lithification (cementing pieces together) of sediment	shale, sandstone
Metamorphic	rocks subjected to high pressures and temperatures	Recrystallization in solid state (rearrangement of atoms without melting)	gneiss, slate

Figure 10.1: Different Rock Types Form in Different Ways

Igneous rocks form when molten, liquid rock (lava or magma) cools and solidifies. To get lava you need a volcano, and to get a volcano you need a hot, partially molten interior. Thus the presence of igneous rocks on a planet implies a hot molten interior when the rocks formed. All of the terrestrial planets started out with hot, partially molten interiors initially, so igneous rocks can be found on all of them.

Sedimentary rocks form when rocks are broken apart into pieces and then essentially glued back together again (this is known as lithification). The "glue" comes from minerals precipitated by water. Thus the presence of sedimentary rocks on a planet implies liquid water on the surface when the rocks formed. Sedimentary rocks can be found on the Earth and Mars, implying that though the Martian surface is currently cold and dry, liquid water once ran across it.

Metamorphic rocks form when rocks are subjected to high temperatures and pressures. For this to happen on a significant scale at a planet's surface, you need plate tectonics. Thus the presence of metamorphic rocks on a planet implies plate tectonics on the surface when the rocks formed. The only planet to have or had plate tectonics that we know of is the Earth, so metamorphic rocks are expected only on the Earth's surface.

Now what if we want to study the interior of a planet? How can we do this? One option you might suggest is drilling; if we want to know what the inside of the Earth looks like, why don't we just drill a deep hole? The Russians had a project from 1970-1989 to do just this. However, in 19 years of drilling, they only got down 12 km (or 7.5 miles). This might seem deep...until you remember that the Earth is about 6,700 km to the center! Thus, 12 km is only 0.18% of the way to the center. In other words, if the Earth were an apple, we haven't even drilled past the skin.

Figure 10.2: Xenoliths (the light chunks) (https://commons.wikimedia.org/wiki/File:Peridotite_xenoliths2.tif

Another suggestion that you might come up with is to look at rocks spewed up by volcanoes from the Earth's interior. Such rocks are called xenoliths (Figure 10.2), which literally means "strange rock" in Latin. However, it turns out that the deepest xenoliths come from depths of about 300 km. This is better than before: xenoliths can tell us what the layer just under the surface looks like, but it's still not great. 300 km is still only 4.5% of the way to the center. Thus if we want to know what the center of the Earth is like, we are going to need to be more creative.

One option is to look at meteorites. Meteorites are extraterrestrial material that falls to Earth. Most meteorites are from the asteroid belt; they are the leftover material from which the Earth formed. Thus by looking at meteorites, we can get a general idea of what elements make up the Earth. It's like looking at a messy kitchen after someone makes a cake to take a guess at what's in the cake (flour on the floor, egg shells in the sink, so on...). What we learn when we look at meteorites is that there is a lot more metal (notably iron) in them than we see at the Earth's surface. This suggests that somewhere in the Earth, there is significant amounts of metal.

The fact that the Earth has a magnetic field further suggests that this metal is liquid. This is because we know from physics that to create a magnetic field, you need liquid, moving metal.

We can use the overall average density of the Earth to estimate the amount of metal. The density of an object is its mass divided by its volume. You can think of the density of an object as how compact the atoms that make up the object are. If the atoms are more compact, the object is denser. Keep in mind that density is not the same thing as weight. A large hunk of iron and a small hunk of iron will have the same density because the atoms making up the hunks are equally compact, but the large hunk will weigh more because there is more of it. Thus density does not depend on the size of something, but only on what something is made of. In general, ice and gas have a density of about 1000 kg/m^3, rock has a density of about 3000 kg/m^3, and metal has a density of about 8000 kg/m^3. So if a planet were made entirely of metal, it would have a density of 8000 kg/m^3. If a planet were made entirely of rock, it would have a density of 3000 kg/m^3. If a planet were made of exactly half metal and half rock, it would have a density of 5500 kg/m^3 (exactly in between 3000 and 8000). The average density of the Earth is 4070 kg/m^3, in between these two numbers. This indicates that the Earth is made of both rock and metal, with more rock than metal. As another example, Jupiter has an average density of 1300 kg/m^3. This indicates that Jupiter is made mostly of ice and gas with a little bit of rock.

So far, we have talked about several methods which give us some information about planets' interiors. However, we have not yet talked about the most significant one for the Earth. Most of our information about the Earth's interior comes from studying seismic waves, or earthquake waves.

Earthquakes are a sudden release of energy and the subsequent shaking of the ground. Any event that releases a large amount of energy all at once can cause an earthquake - things like faulting, volcanic eruptions, bomb blasts, meteorite impacts, and landslides. Most of the Earth's earthquakes are created by faulting due to plate tectonics.

Once an earthquake occurs, its energy is radiated away from the origin of the earthquake in the form of seismic waves, which are waves that shake the ground that they travel though. This is analogous to dropping a pebble in a pond and watching the waves ripple outward in all directions from the drop.

Seismic waves can be measured using seismometers, which are instruments that record the motion of the ground over time. The ground motion is recorded in a series of squiggles known as a seismogram.

A typical seismogram will show three distinct sets of waves (Figure 10.3). The first set is known as the P-wave (or Primary wave), the second set is known as the S-wave (or Secondary wave) and the final set is known as the Surface waves. P and S waves are both types of body waves, meaning they travel through the Earth's interior. In contrast, surface waves can only travel through layers near the Earth's surface. Since we are interested in the Earth's interior, we are going to focus on the body waves.

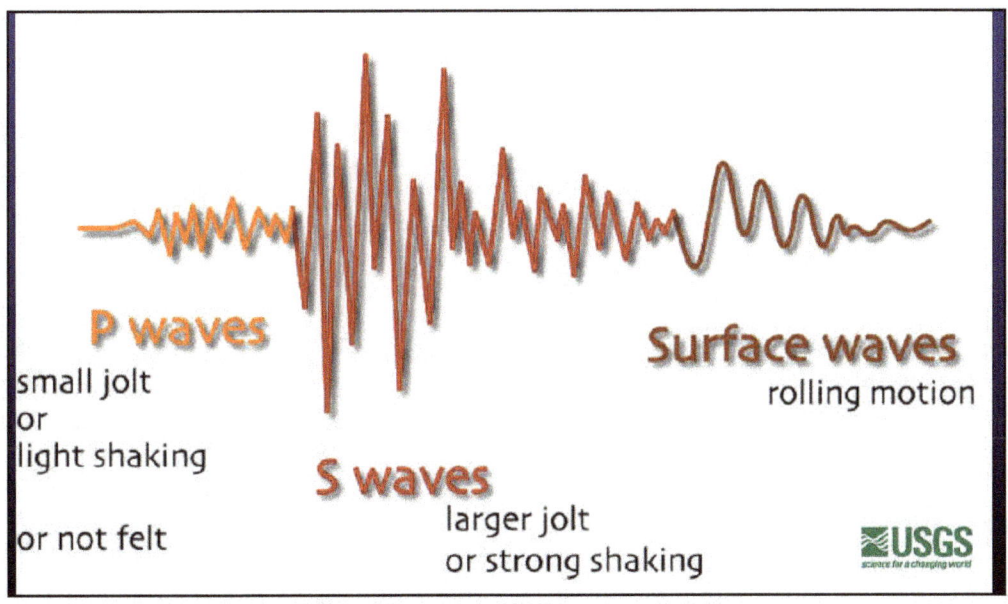

Figure 10.3: Typical Seismogram (USGS)

The two types of body waves differ in several important ways. P-waves push and pull the rock that they travel through, S-waves shear it side-to-side or up-and-down (Figure 10.4). P waves travel faster than S-waves. Lastly, P-waves can travel through solid and liquid, but S-waves can only travel through solids.

By observing the speeds and patterns of body waves as they travel through the Earth, scientists can learn about the Earth's interior. For example, let's suppose an earthquake occurs at the star in Figure 10.5. The paths of S-waves from this earthquake are shown by the arrows. Note that Seismic stations within 103 degrees of the source will observe both P and S waves from this earthquake.

However, notice that when S-waves try to go through the core, they get blocked. This means that seismic stations on the opposite side of the Earth will observe only P-waves. No S-waves will be recorded; this is known as the S-wave shadow zone (gray hashed region in Figure 10.5). Recall that S-waves can't travel through liquid. Thus, the S-wave shadow zone tells scientists that the core of the Earth must be liquid. The liquid core blocks S-waves from reaching the shadow zone. The size of the core can be constrained by the size of the shadow zone. The bigger the shadow zone, the bigger the core must be.

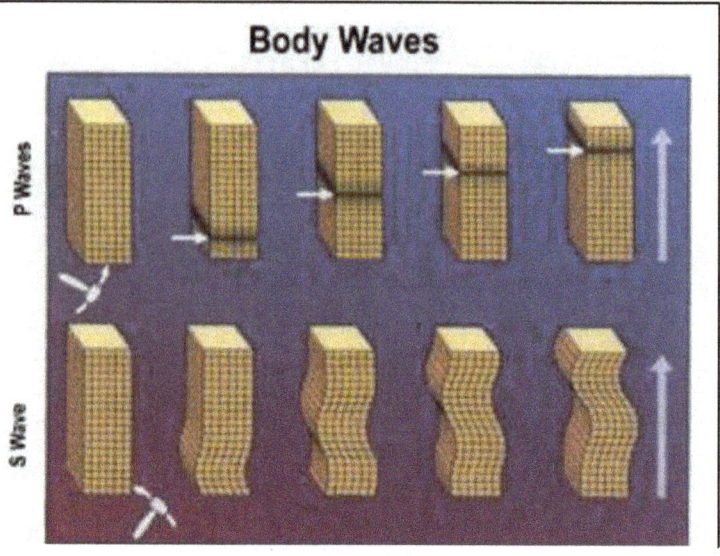

Figure 10.4: P and S Waves
(https://upload.wikimedia.org/wikipedia/commons/5/5c/P swaves_ro.jpg)

We can sum up the structure of the Earth into three compositional layers (Figure 10.6). The outer layer, the crust, ranges from 0-60 km thick and is composed of low density rocks. The crust is very thin relative to the Earth as whole; a good comparison is the skin of an apple relative to the whole apple. The middle layer, the mantle, is composed of high density rocks and goes from below the crust to 2900 km deep. At the center is the core, which is made of metal, primarily iron (95% iron and 5% nickel). There are two parts to the core; the outer core and the inner core. The outer core is liquid, and the inner core is solid. This might seem counterintuitive to you; if the Earth is hot enough to melt iron at the outer core, why not the inner core as well? The answer is that melting temperature depends on pressure. Though the inner core is hotter than the outer core, the pressure is higher, making it solid.

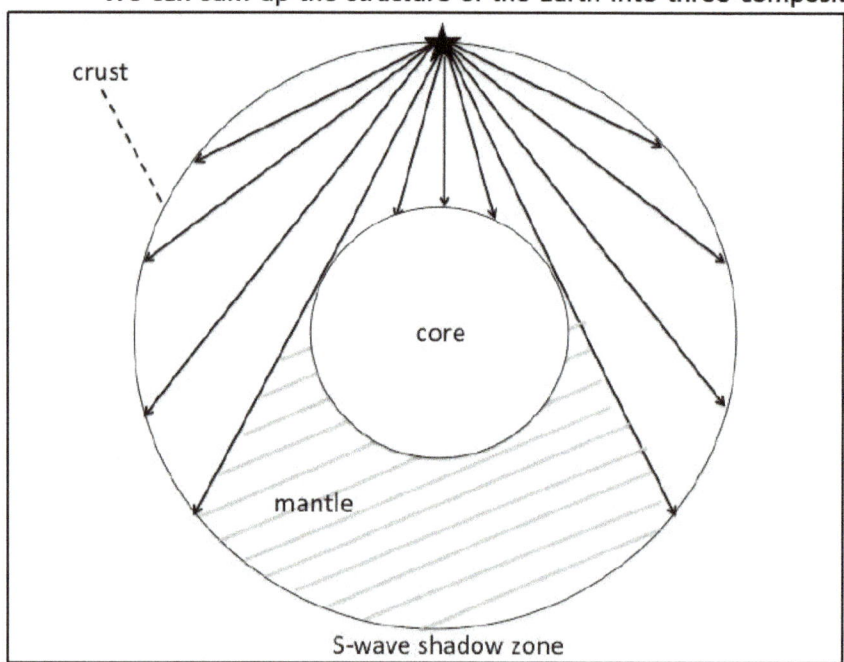

Figure 10.5: S Wave Shadow Zone

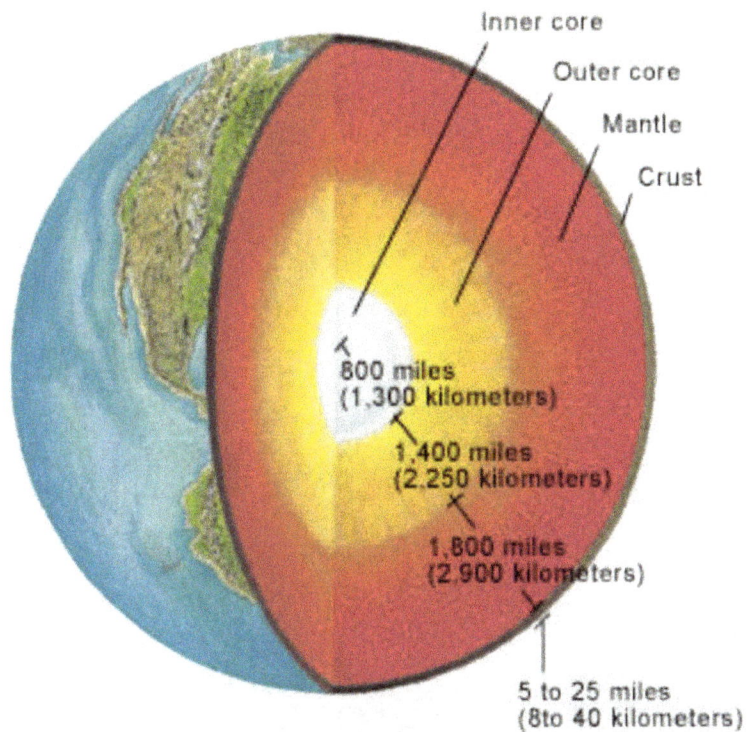

Figure 8.2: Layers of the Earth
(https://commons.wikimedia.org/wiki/File:Earth_layers_NASA.png)

Questions You Should Be Able to Answer
1. What are the three different types of rocks? How does each type form? What does each type tell us about the planet it is found on?
2. How do scientists determine the interior structure of the Earth?
3. What is density? How can density be used to determine what the inside of a planet is like?
4. What is an earthquake?
5. What causes earthquakes?
6. What is a seismic wave?
7. What are the two types of seismic waves? What is the difference between them?
8. What are the two types of body waves? What is the difference between them?
9. How do scientists use seismic waves to determine the interior structure of the Earth?
10. How do we know that the Earth's outer core is liquid?
11. Describe the following layers of the Earth: crust, mantle, outer core, inner core.

Suggestions for Further Reading
1. http://ratw.asu.edu/aboutrocks.html (about rocks)
2. http://www.khanacademy.org/science/cosmology-and-astronomy/earth-history-topic/seismic-waves-tutorial/v/seismic-waves (Khan Academy explains seismic waves)

Chapter 11: Crater Dating

In this chapter, you will learn how scientists use craters to determine the ages of planetary surfaces.

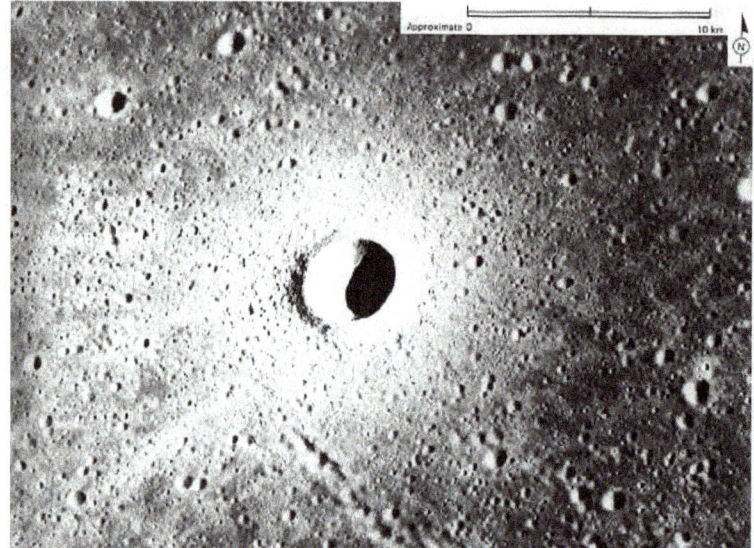

Figure 11.1: Typical Crater on the Moon (http://www.nasa.gov, Apollo 17)

An impact crater forms when solar system debris (typically from the asteroid belt, but sometimes a comet) hits the surface of a planet or moon. The collision creates an explosion and leaves behind a bowl-shaped depression, or crater. Around the crater there will be debris that was thrown out during the impact; this is known as the ejecta blanket. A typical impact crater on the moon is shown in Figure 11.1; the ejecta blanket is the lighter colored material surrounding the crater.

The size of the crater that forms depends on the gravity of the planet or moon that is hit, the size of the impactor, and the speed that the impactor is traveling. However, to give you some idea, here are crater sizes for typical different sized impactors on the Earth. A 10 ft in diameter meteor will create a crater that is 620 feet across - this is 2 football fields! A 100 ft meteor gives you a crater that is 3/4 of a mile across, a 1,000 ft meteor gives you a crater that is 5 miles across, and a 10 mile meteor gives you a crater that is 200 miles across (the later is the size of the meteor that caused the extinction of the dinosaurs 65 million years ago).

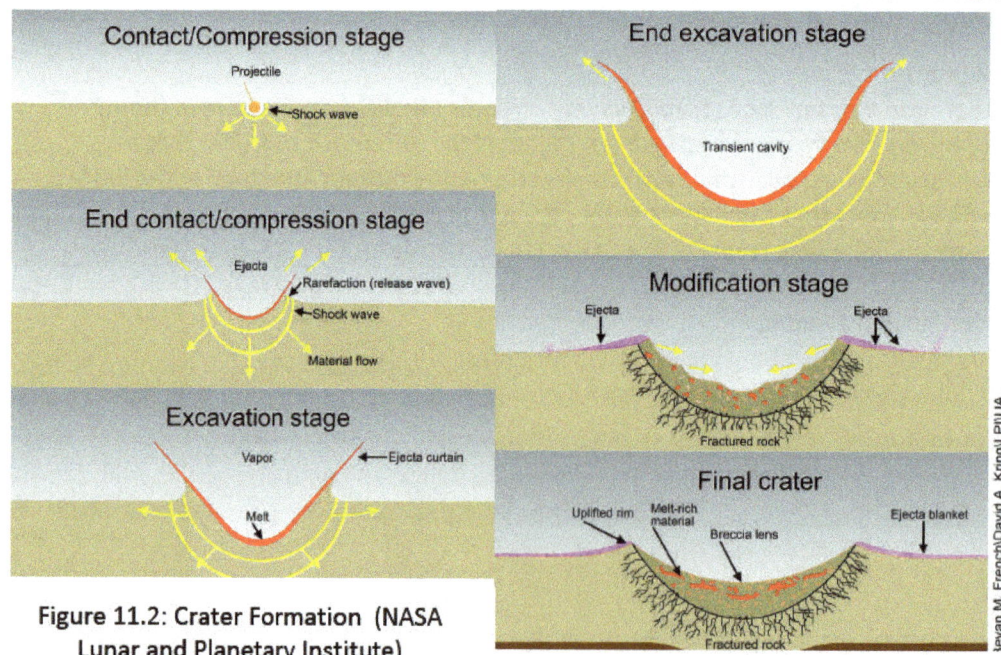

Figure 11.2: Crater Formation (NASA Lunar and Planetary Institute)

Figure 11.2 illustrates crater formation. A meteor comes in, and embeds itself into the surface. The energy from the impact sets up an explosion very similar to a nuclear bomb being detonated. The explosion throws material upward and outward, creating the crater and the ejecta blanket. Most of the meteor itself is vaporized due to the extremely high pressures and temperatures.

Craters can give scientists a sequence of events by looking at overlapping relationships. If two craters overlap, then the crater on the top is younger than the crater on the bottom (Figure 11.3a). In other words, the crater on the bottom happened first and the crater on top happened second. Visible ejecta rays form at the same time as the crater, and so are included in the rule. For example, in Figure 11.3b, the crater with rays is younger than the crater without rays. As another example, in Figure 11.3c, the crater with rays is older than the crater without rays. If two craters don't overlap then we can't say which is older (Figure 11.3d).

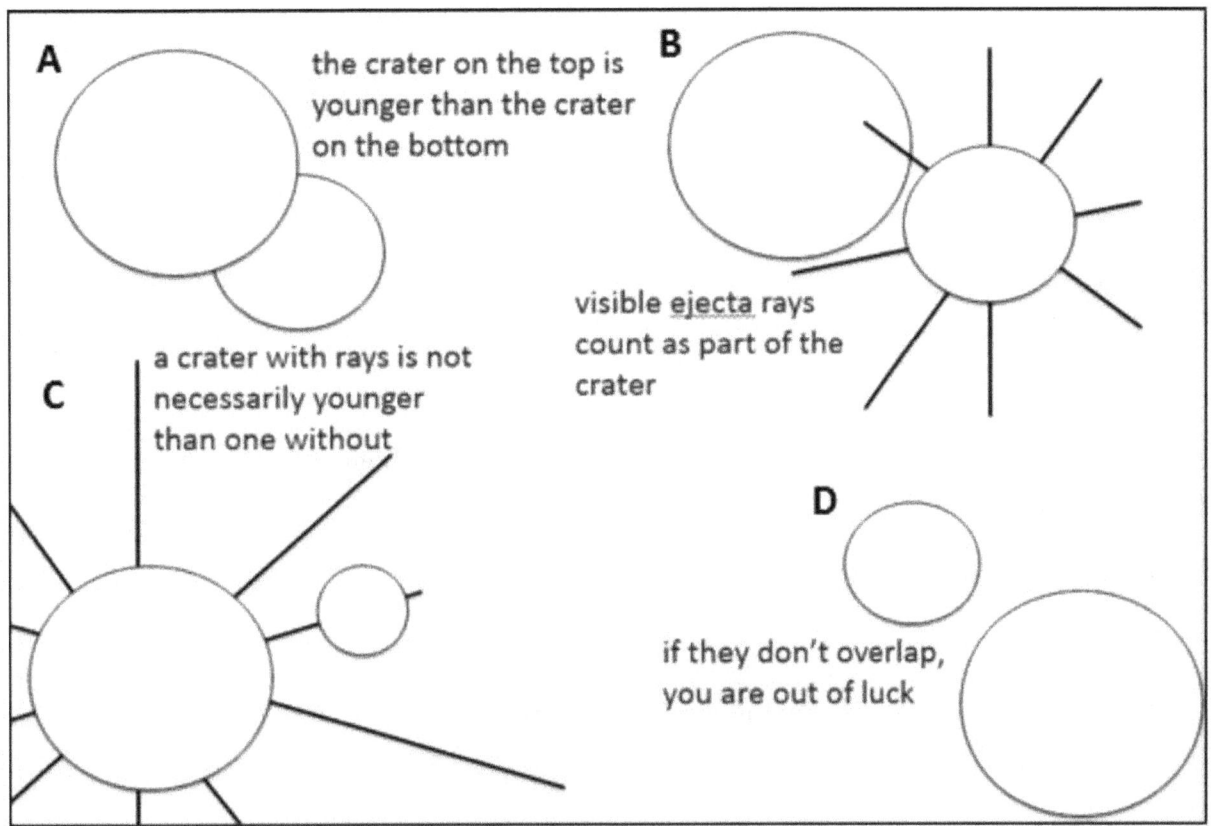

Figure 11.3: Relative Age Dating with Craters

Sometimes craters have been filled in with lava (which typically cools to the igneous rock basalt). If this is the case, the craters that are filled in with lava came before the lava, and the craters that are not filled in with lava came after the lava. In Figure 11.4a, the bottom crater formed, then the crater was filled in with lava, and then the top crater formed. In Figure 11.4b, both craters are infilled with lava, so the lava must have come last. Thus, first the unrayed crater formed, then the rayed crater, and then both were filled with lava. Note that if two craters don't overlap, but one is filled with lava, we can now get a sequence of events (Figure 11.4d). The crater that is filled with lava must be older because it has to come before the lava and the crater that is not filled with lava must be younger because it has to come after the lava.

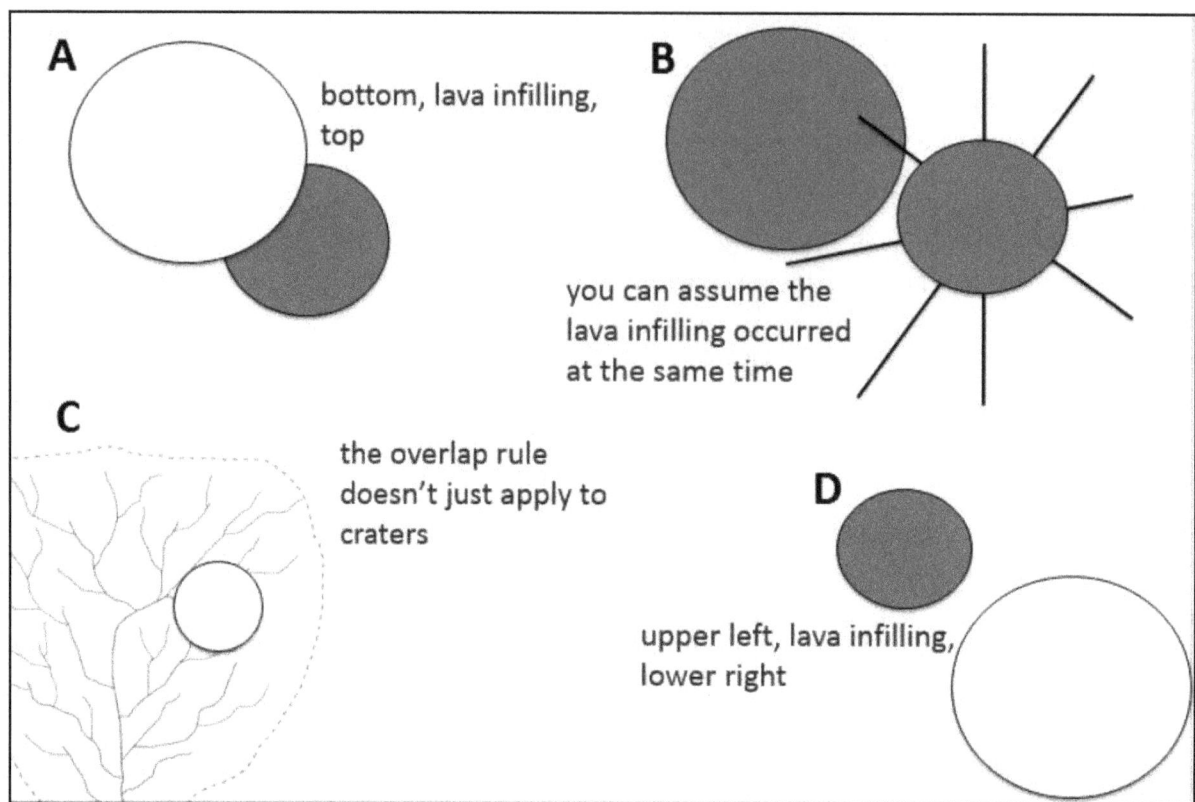

Figure 11.4: Relative Age Dating with Craters, continued

Not only do craters give us a sequence of events, but the number of craters tells us something about a surface's age. The age of a surface refers to the last time that it was significantly changed. This can be via volcanism, wind and/or water activity, or plate tectonics. Basically anything that erases the surface and forms or exposes new rock counts. If a scientist tells you than a certain area of Mars is 100 million years old, she means that it's been 100 million years since anything significant like this happened.

To see how this works, let's imagine two areas on the same planet. Since they are on the same planet, they experience the same cratering rate. On average, each time a crater forms on one surface, a crater forms on the other surface. Thus over time the surfaces accumulate the same number of craters. As time goes on, more and more craters are added.

But then something happens that erases all of the craters on one of the surfaces...maybe volcanism, maybe plate tectonics, it doesn't matter what. Now this surface has a clean slate. As time continues to go on, both surfaces will continue to accumulate craters at the same rate, but the surface that was erased will always have less craters because it was modified more recently.

We can generalize this into the observation that the number of craters gives a surface's relative age. The more craters a surface has, the older it is. The fewer craters a surface has, the younger it is. In other words, the surface with fewer craters was modified more recently.

If you know the cratering rate, then you can get the actual numerical age. For example, suppose the cratering rate is 1 crater every 1 million years. This means that a surface with 9 craters is 9 million years old and a surface with 3 craters is 3 million years old.

The difficulty with this is that rock samples are needed to constrain the cratering rate (through radiometric dating). However, once the cratering rate is known, you can go back and determine the age anywhere on the planet. For example, let's suppose you send an astronaut to a region of the moon with 9 craters and he brings back rock samples. You use radiometric dating to determine the age of the rocks at 3 million years. This tells you that the cratering rate is 3 craters every 1 million years (9 craters divided by 3 million years). You can then use that rate to determine the age any other surface on the moon. For example, a surface with 3 craters must be 1 million years old since 3 craters form every 1 million years.

Unfortunately, the only solar system body for which we have samples of known origin is the Earth's moon. So the cratering rate is only definitively known for the Moon. However, scientists estimate the cratering rate for other bodies (like Mars or Venus) by taking the Moon's cratering rate and adjusting for size (bigger bodies have more gravity and so pull in more objects) and location in the solar system (the closer you are to the asteroid belt, the more cratering there will be). Lastly, it should be noted that the cratering rate is not constant; it changes with time. Scientists need to account for this as well when determining surface ages.

Questions You Should Be Able to Answer
1. What is an impact crater? How are impact craters formed?
2. Be able to do relative age dating using impact craters, such as done in class and on the homework.
3. How can I tell the relative age of a surface using craters?
4. What do I need to determine the absolute (numerical age) of a surface using craters?

Suggestions for Practice
1. http://astro.unl.edu/classaction/questions/terrplanets/ca_terrplanets_cratersandage1.html (surface ages questions)
2. http://astro.unl.edu/classaction/questions/terrplanets/ca_terrplanets_cratersandage2.html (more surface ages questions)

Chapter 12: Impact Hazards

In this chapter, you will learn about the hazards that impact events pose. We will discuss the effects of large and small impact events, the frequency with which such events occur, and possible mitigation strategies.

This is a topic that has often has been featured in films, for example Armageddon and Deep Impact. Today we will see just how realistic these films are! It is worth first noting the large impacts do still happen. For example, in July of 1994, the 3-mile in diameter comet Shoemaker-Levy impacted into Jupiter, breaking up into pieces at it did so. The explosions left behind dark scars in the atmosphere and released more energy than the world's total nuclear arsenal combined (Figure 12.1). The dark scars took several months to dissipate. A similar sized impact on the Earth would cause global devastation and the likely extinction of the human species.

Figure 12.1: Scars Left by the Impact of Comet-Shoemaker Levy on Jupiter (http://www.nasa.gov, Hubble)

How do scientists know what happens when a meteor hits the surface? Two ways: by going out into the field and looking at the craters left behind by impacts on the Earth and by using computer simulations.

Let's look first at the effects of a very large impact, such as the impact that formed Chicxulub crater. Chicxulub crater is a 200 mile wide crater on the Yucatan Pensinsula in Mexico. It was created when a 10 mile wide asteroid (this is about twice the height of Mount Everest for reference...!) hit the surface 65 million years ago. Scientists believe that the dinosaurs went extinct because of this impact event.

The Earth experienced several negative effects both on the short term and long term scale after this impact. The energy of the impact immediately triggered earthquakes and tsunamis, and a heated vapor plume (essentially a fireball of thermal radiation). The vapor plume and hot ejecta falling back to the Earth led to massive, global forest fires. In the months to years following the impact, the dust and soot thrown out by the impact and subsequent fires blocked photosynthesis, leading to an impact winter and a collapse of the food chain. It is estimated that the dust and soot would have made it too dark to see for 1 to 6 months, and that photosynthesis would have been blocked for 2 months to a year. The dust and soot eventually fell back to the Earth, creating a global rock layer known to geologists as the K-T Boundary. In the long term, carbon dioxide built up in the atmosphere (because there were no more plants to remove it), leading to global warming. The ozone was destroyed from sulfate aerosol formation. It took the Earth's centuries to recover from these long term effects. In the end, somewhere between 50 and 80% of all species living at that time went extinct.

Even small impacts can cause devastation on a local scale. An example is the Tunguska event. On June 30, 1908 at 7:17 am, a 50 meter meteorite exploded 4 miles above the ground in Tunguska, Siberia. No one was near enough to be injured, but observers reported a "deafining bang" and a fiery cloud. The shock wave from the impact broke windows and knocked people off their feet miles away. When people investigated the area later, they found a 25 mile wide area where trees had been burned and flattened radially by the explosion (Figure 12.2). This is about the same size as Washington D.C.

Figure 12.2: Flattened Trees at Tunguska (https://commons.wikimedia.org/wiki/File:Tunguska_e vent_fallen_trees.jpg)

If the meteor at Tunguska had hit the surface, it would have created a crater about 3/4 of a mile in diameter. This is about the size of Meteor Crater in Arizona. Meteor Crater was formed 50,000 years ago, in an area that was fortunately probably not inhabited by humans at the time.

However, computer simulations tell us what humans would have experienced had they been there. A fireball would have burned anything within a radius of about 6 miles from the impact. Within 15 miles, large animals would have been killed or wounded from the shockwave. Hurricane force winds would have been experienced up to 25 miles away.

To put this into perspective, let's imagine that such an impact happens in Times Square instead of the Arizona desert (Figure 12.3). A shows the zone that gets burned from the fireball, B shows the zone where large animals are killed or wounded from the shockwave, and C shows the zone that experiences hurricane force winds. The star represents the location of Westchester Community College in Valhalla, NY.

63

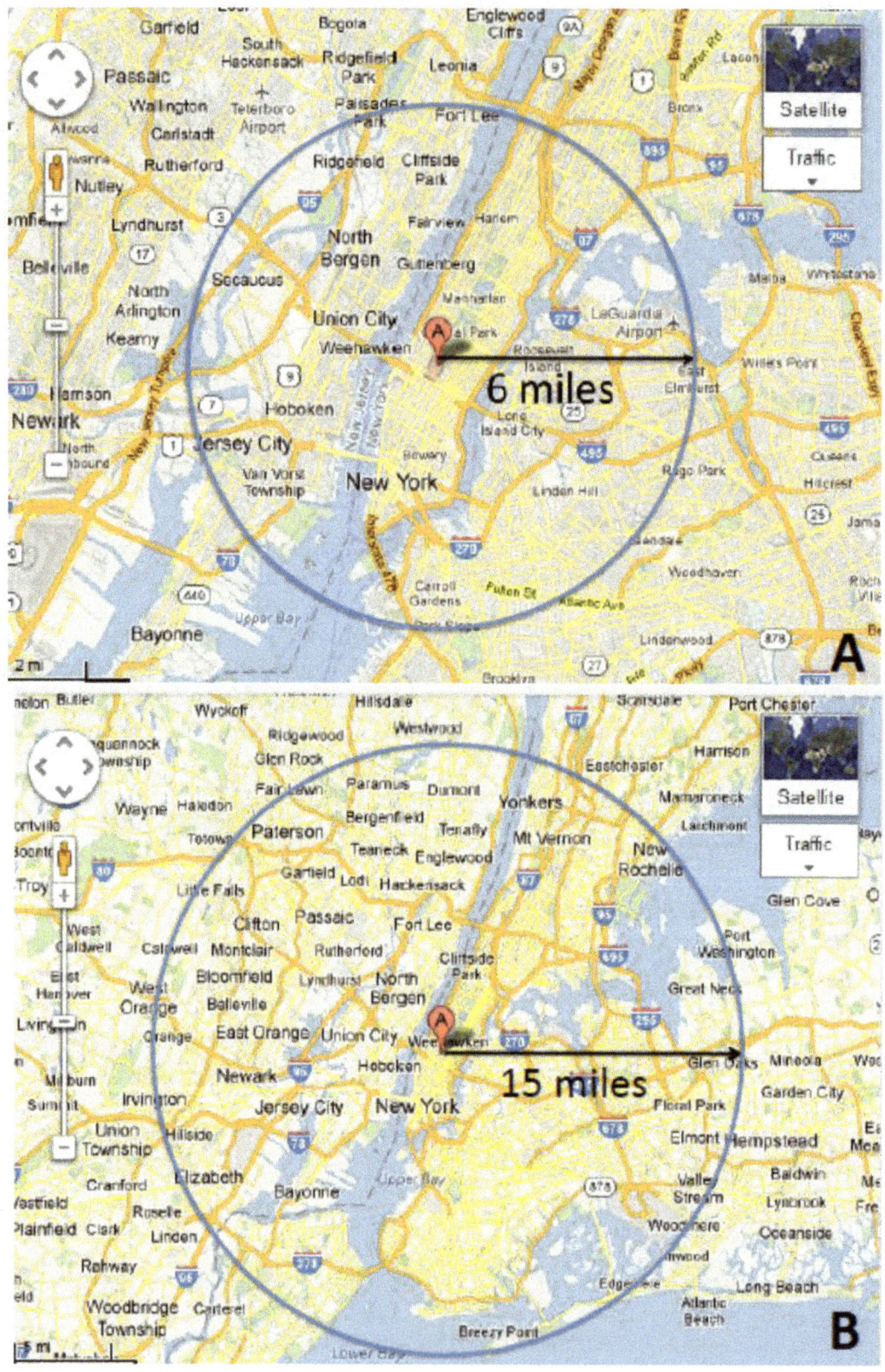

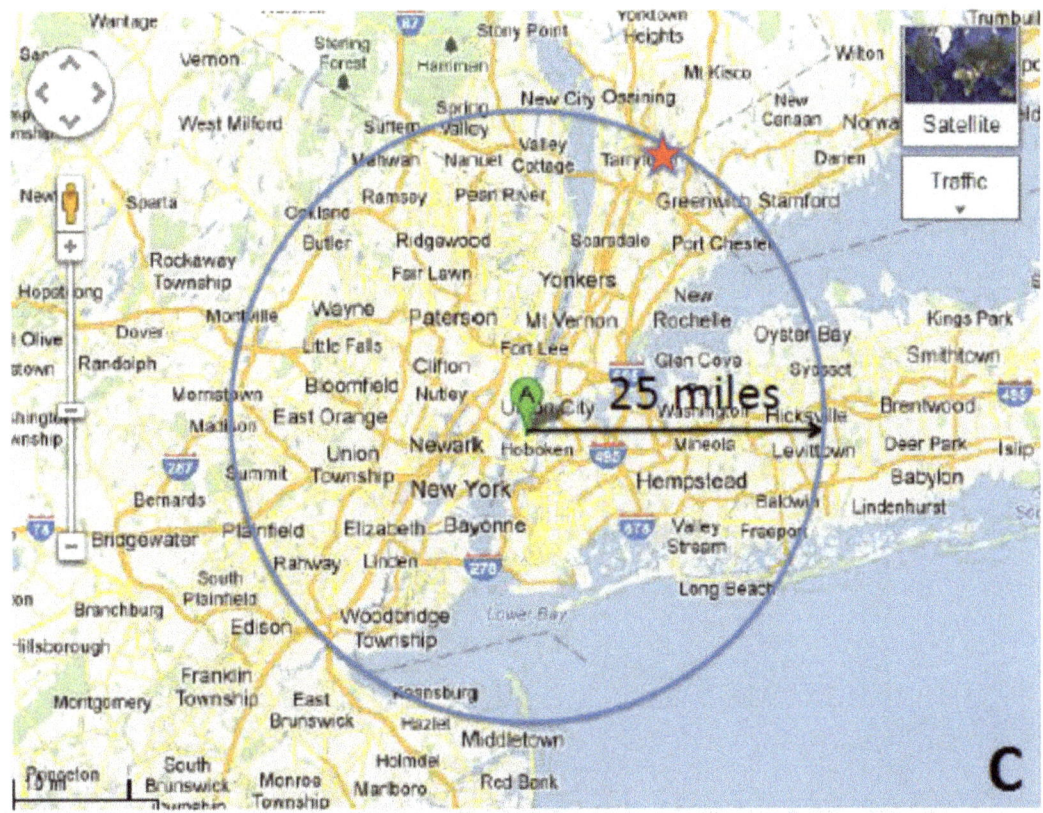

Figure 12.3: Effect of a 50-meter Impactor Hitting Times Square, New York City. Map data: Google

We saw that large impacts can cause mass extinctions and global devastation. We also saw that even small impacts can cause a lot of devastation if they hit in the wrong place. Our next question is how often do these events happen? The plot in Figure 12.4 tells us. On the x-axis we have the size of the impact and the corresponding impact energy. Note the scales are logarithmic. On the y-axis we have the recurrence interval. This is the average number of years between impacts of that size. For example, an impact with a recurrence interval of 100 years occurs on average once every 100 years. This does not mean that an impact occurs every 100 years like clockwork; it means that on average in a period of 100 years there was probably one impact. There might have been none, or two, or three, but most likely there was one. Another way to think about this is that each year there is a 1/100, or 1%, chance of an impact this size happening. Reading the chart, we see for example that an impact the size of Chicxulub occurs on average once every 50 million years. This means that each year there is a 1/50 million, or 0.000002%, chance of an impact the size of Chicxulub occurring.

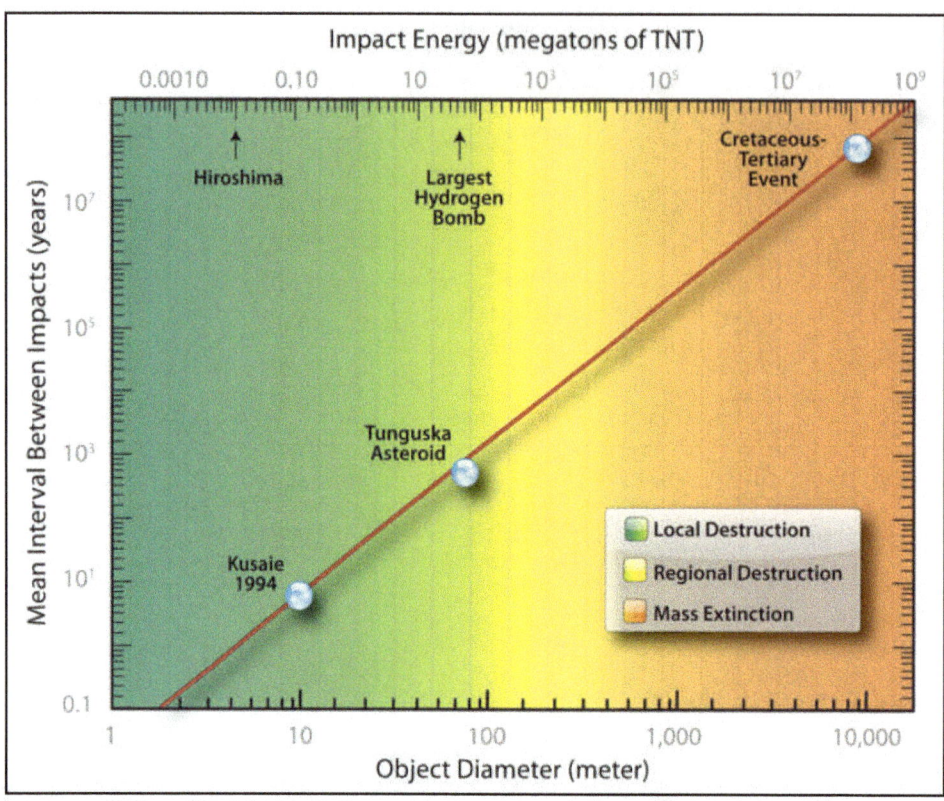

Figure 12.4: Impact Size versus Frequency (http://www.lsst.org/)

An NEO, or Near Earth Object, is an asteroid or comet with an orbit that intersects the Earth's orbit, or passes very near the Earth's orbit, and hence has the potential to hit the Earth. The orbits of all known NEOs (as of early 2013) are is shown in Figure 12.5. The orbits of Mercury, Venus, Earth, and Mars are labeled; all other orbits are the orbits of NEOs. As you can see, there are quite a lot.

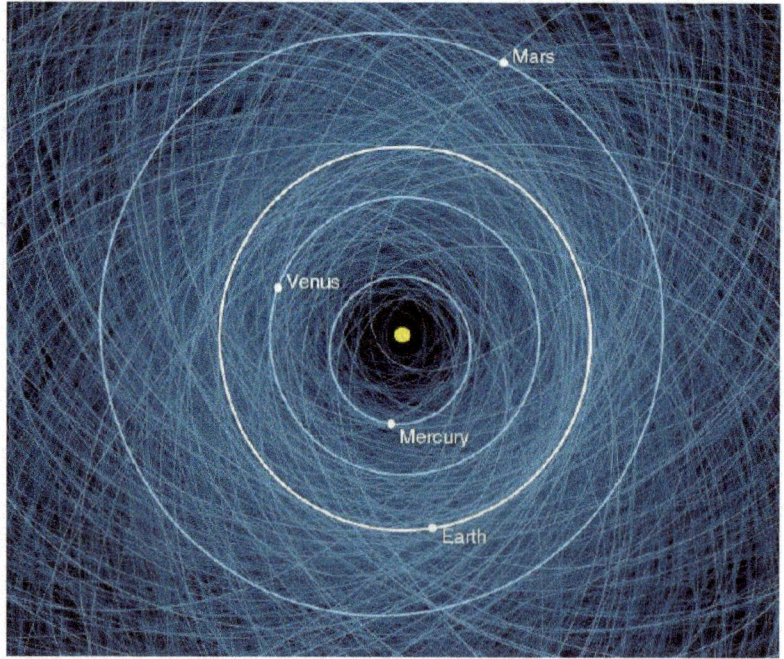

Figure 12.5: The Orbits of Near Earth Objects (NEOs) (http://www.nasa.gov)

Most of these were detected as part of NASA's NEO program. The initial goal of this program was to detect at least 90% of NEOs bigger than 1 km (the threshold for mass extinction) by 2008. So far (as of June 5, 2018) 18,214 NEOS have been discovered. 893 of them have a diameter bigger than 1 km. Since it is estimated there are about 1,000 NEOs with diameters bigger than 1 km, this is almost 90%. Thus the new goal of the program is to detect at least 90% of NEOs greater than 140 m by 2020. So far 8,172 of these have been found.

Which leads us to the question – what do we do if we discover an NEO headed towards us?

The National Research Council recently put together a committee of scientists to analyze this question and create a report (Figure 12.7). They recommended three ways to deflect an asteroid.

The first method is known as the gravity tractor method. In this method, the gravity of a spaceship is used to slowly pull the asteroid off of its current trajectory. It's like a tractor, but the spaceship is not attached. The benefit of this method is that it is very precise. The disadvantage is that the gravitational force between the asteroid and the spaceship is very small, so this method only works if you have decades of time and a relatively small asteroid.

If you have less time or a larger asteroid, you can use a kinetic impactor, where you try to deflect the asteroid by hitting it with something. This is not as precise as the gravity tractor, and there is always the chance that you may break up the impactor into several pieces instead of deflecting it!

For the largest objects, a kinetic impactor will not provide enough energy, and a nuclear denotation will be needed.

Thus actions may involve studying and monitoring the asteroid, a characterization mission to the asteroid, agreements between nations, sheltering and evacuation, the gravity tractor, the kinetic impactor, and the nuclear detonation. The appropriate actions depend on the scale of the event and how much warning time we have. For the smallest scale events, the best action is sheltering and evacuation. For medium and large scale events, if there is little time, sheltering and evacuation is the only option. However, for longer warning times we could try to deflect the asteroid using one of the methods we talked about. The committee also found that some events are so large that there would be inevitable global catastrophe that we don't have the capability to avoid. As you can see, the committee labeled these events as Earths with x's over them.

67

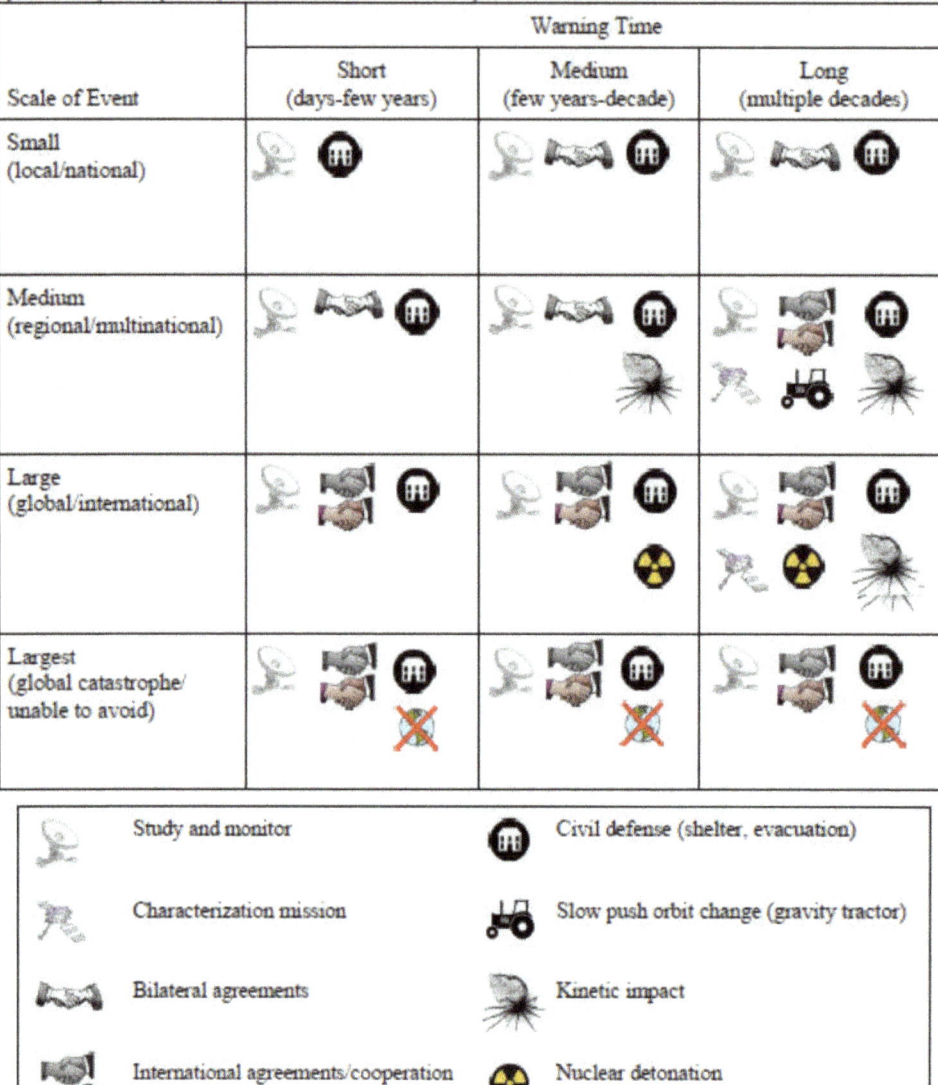

Figure 12.7: Impact Mitigation Strategies (NRC)

Questions You Should Be Able to Answer
1. Describe some of the effects of large and small(er) impact events.
2. What is a recurrence interval?
3. Do large impacts happen more or less frequently than small impacts?
4. What is a NEO? What is the NEO program and what are its goals?
5. What are some of the ways scientists are working on to mitigate the effects of a large impact event?

Suggestions for Further Reading
1. http://neo.jpl.nasa.gov/ (NASA's Near Earth Object Program website)
2. http://www.pbs.org/wgbh/nova/earth/meteor-strike.html (PBS Nova program about the Chelyabinsk Meteor)

Chapter 13: The Moon

In this chapter, you will learn about the Moon. We will talk about the exploration, evolution, and formation of the Moon.

Missions to the moon (US)

- Pioneers (1958 and 1959): flybys
- Rangers (1961-1965): crash landers
- Surveyors (1966-1968): soft landers
- Apollo Program (1968-1972): manned missions
- Clementine (1994): orbiter
- Lunar Prospector (1998): orbiter
- Lunar Reconnaissance Orbiter, LRO (2009): orbiter
- Lunar Crater Observation and Sensing Satellite, LCROSS (2009): impactor

Figure 13.1: Missions to the Moon

Missions to the Moon (Figure 13.1) have mostly focused on preparation for the manned lunar missions, the Apollo Program. Since then, there have only been 4 US missions to the Moon.

The Moon is smaller than the Earth; its mass is about 1% the Earth's mass and its radius is about 27% the Earth's radius. Additionally, the Moon's density is significantly smaller than the Earth's density; this implies that the Moon has proportionally less metal than the Earth does. Figure 13.2 show two sides of the Moon; on the left is the side that always faces the Earth (the near side) and on the right is the side that always faces away from the Earth (the far slide). Note that there appears to be two major terrains on the Moon; the darker areas, and the lighter areas.

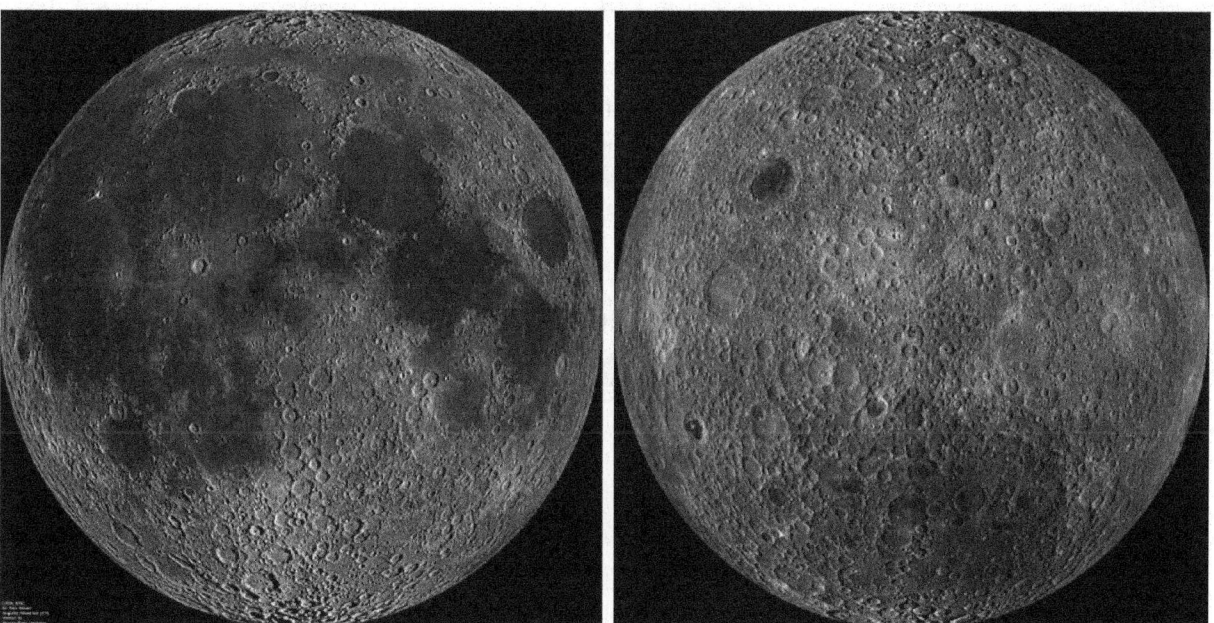

Figure 13.2: The Moon (http://www.nasa.gov)

If we look at a topographic map of the Moon (Figure 13.3), we can see that the darker areas generally coincide with areas that are lower in elevation. Hence we call them the lowlands, or the maria. The lighter areas generally coincide with areas that are higher in elevation. Hence we call them the highlands.

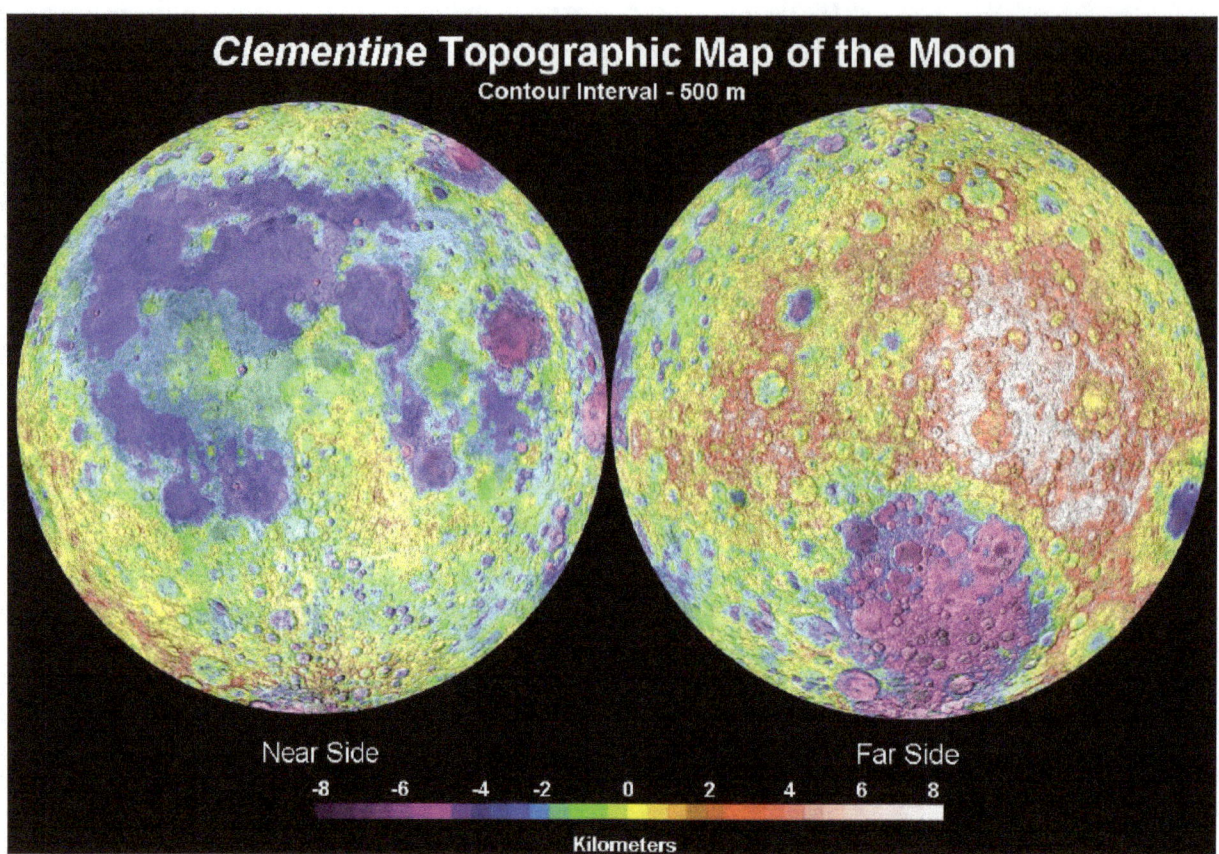

Figure 13.3: Topography of the Moon (http://www.nasa.gov, Clementine)

The lowlands have less craters than the highlands, indicating that they are younger. The lowlands are composed of low lying, smooth lava flows, while the highlands are composed of jumbled mountains that were pushed up by large impact cratering events. The lowlands are composed of the igneous rock basalt, which is dark in color. The highlands are composed of the igneous rock anorthosite, which is light in color.

We can put this information together to come up with a sequence for the evolution of the Moon. When the moon first formed 4.6 billion years ago, the surface was liquid rock; we call this a magma ocean. From 4.6 to 4.1 billion years ago, differentiation occurred in this magma ocean. Heavy minerals sank to the bottom and lighter minerals rose to the surface. The minerals that rose to the surface solidified, forming the highlands.

This was followed by a period of late heavy bombardment from 4.0 to 3.8 billion years ago. During this time, the cratering rate on the Moon spiked for unknown reasons (Figure 13.4). Giant impact basins, like the Orientale Basin, which is as big as Texas, were formed.

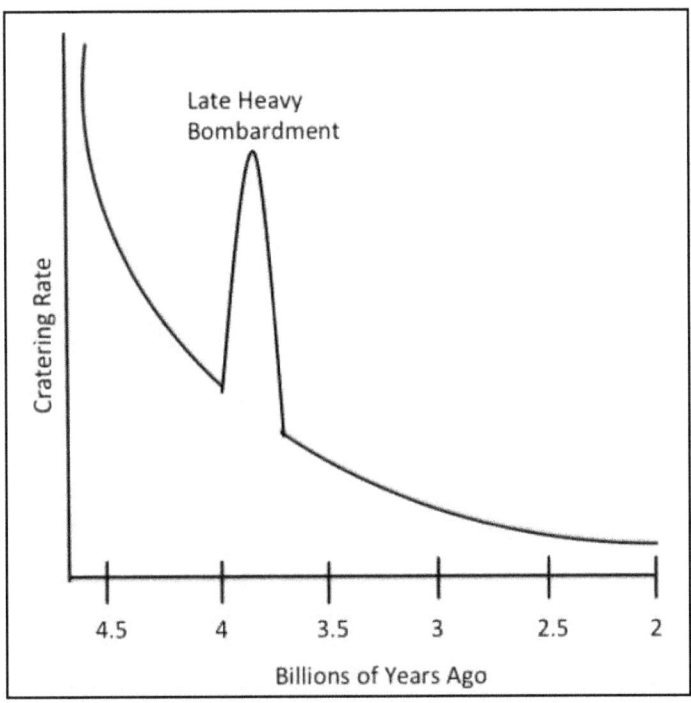

Figure 13.4: Late Heavy Bombardment

The giant impact basins formed during the late heavy bombardment weakened and cracked the crust. From 4 to 2 billion years ago, lava rose up through these cracks and pooled in the impact basins. When the lava cooled, it formed the lowlands.

By about 2 billion years ago, the Moon's interior had cooled sufficiently that the interior was no longer partially molten and volcanism stopped. Since then, the moon has been geologically dead. The only activity now is cratering.

Where did the moon come from in the first place? Following the Apollo missions to the Moon, scientists developed several hypotheses to answer this question. We will discuss the four most popular at the time. The binary accretion hypothesis suggests that the Moon and the Earth formed at the same time from the same cloud of debris. In other words, as a cloud of debris coming together formed the Earth, the outer edges of that cloud came together to form the Moon via the same process. The capture hypothesis suggests that the Moon is an asteroid that drifted too close to the Earth and was captured by the Earth's gravity. The fission hypothesis suggests that the Moon formed when the Earth broke up because it was spinning too fast. Lastly, the giant impact hypothesis suggests that the Moon was formed from material thrown off during a large impact event.

When scientists have several hypotheses, they analyze the available evidence to see which hypothesis is most consistent with that evidence. We will do this together in class to figure out which hypothesis makes the most sense!

Questions You Should Be Able to Answer
1. How does the moon's mass, radius, and density compare to the Earth's?
2. What are the differences between the lunar highlands and lowlands?
3. Describe the evolution of the moon from magma ocean to present day.

4. What are the binary accretion, capture, fission, and giant impact hypotheses for the formation of the moon? Which hypothesis do scientists currently favor? What are the problems with the other hypotheses?

Suggestions for Further Reading
1. http://nineplanets.org/luna.html (Nine Planets Moon page)
2. http://nssdc.gsfc.nasa.gov/planetary/planets/moonpage.html (NASA Moon page)

Chapter 14: Mercury

In this chapter, you will learn about Mercury. We will talk about the exploration, surface and interior, and history of Mercury.

There have been very few missions to Mercury; one flyby in 1973 (Mariner 10), and more recently, an orbiter that is currently actively collecting data (MESSENGER - Mercury Surface, Space, Environment, Geochemistry, and Ranging Mission).

Mercury is similar in size to the Moon; its mass is about 6% the Earth's mass and its radius is about 38% the Earth's radius. Mercury's density is significantly larger than the Earth's density; this implies that Mercury has proportionally more metal than the Earth does. Figure 14.1 shows a view of Mercury as mapped by MESSENGER.

Figure 14.1: Mercury (http://www.nasa.gov, MESSENGER)

Mercury is a cratered, airless, and geologically dead world, similar to the Earth's Moon. However, there are some difference between Mercury and the Moon. On the Moon, there are darker areas (the maria, or lowlands). On Mercury, there are terrains of different ages, but they are all the same color (i.e. made of the same rocks). Temperatures on Mercury vary more greatly than on the Moon, from 800 degrees Fahrenheit in sunlight to -280 degrees Fahrenheit in shadow or night. Lastly, Mercury is 60% denser than the Moon, indicating that it has a lot more metal.

There are three basic terrains on Mercury (Figure 14.2). The oldest is the old cratered terrain, which is saturated with craters. Then there is the intercrater terrain, which is less cratered. The youngest terrain is the smooth plains, which has the fewest craters.

Since Mercury is denser than the Earth, this indicates that it has a proportionally larger core than the Earth does. Scientists do not know why this is true; however, our best guess is that a giant impact early on stripped Mercury of some of its rock.

Putting this together gives us a possible history of Mercury. First, a giant impact strips Mercury of some of its lower-

- Old cratered terrain – saturated with craters
- Intercrater terrain – less cratered
- Smooth plains – even less craters

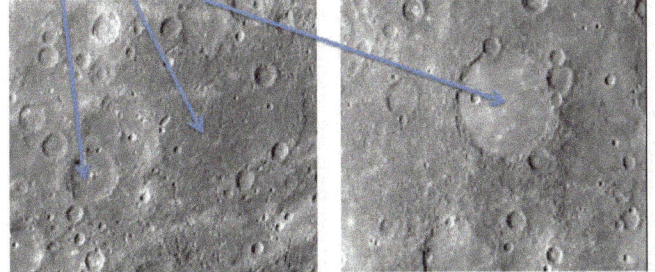

Figure 14.2: Mercury's Terrains (http://www.nasa.gov, MESSENGER)

density rock, creating a small, dense world with a large metallic core. Then the surface cooled, creating the old cratered terrain. Lava flows then created the intercrater plains, and later lava flows created the smooth plains. Lastly, the interior cooled, creating a geologically dead world.

Questions You Should Be Able to Answer
1. How does Mercury's mass, radius, and density compare to the Earth's?
2. How is Mercury similar to and different than the moon?
3. Describe the history of Mercury.
4. What are some of the open (unknown) questions regarding Mercury?

Suggestions for Further Reading
1. http://nineplanets.org/mercury.html (Nine Planets Mercury page)
2. http://nssdc.gsfc.nasa.gov/planetary/planets/mercurypage.html (NASA Mercury page)

Chapter 15: Venus

In this chapter, you will learn about Venus. We will talk about the exploration, atmosphere, surface, and history of Venus.

There have been several missions to Venus (Figure 15.1). Most have been flybys or orbiters. This is because the surface of Venus is too hot to have a lander or rover operate for more than a few minutes before the equipment is fried (as the single Venera lander discovered!).

Missions to Venus

- Mariner 2, 5, and 10 (1962, 1967, 1973): flybys
- Venera (Soviet Union; 1961-1983): flybys, landers, orbiters
- Pioneer Venus (1978-1992): orbiter
- Magellan (1989-1994): orbiter
- Venus Express (ESA; 2005): orbiter
- Akatsuki (Japan; 2010): orbiter

Figure 15.1: Missions to Venus

Venus is similar in size to the Earth; its mass is about 82% the Earth's mass and its radius is about 95% the Earth's radius. Venus' density is comparable to the Earth's density, and Venus is also the closest planet in distance to the Earth. These similarities initially led scientists to call Venus "Earth's Twin." They (incorrectly!) envisioned a Venus much like the Earth - with water and perhaps life. The picture on the left of Figure 15.2 shows you what you would see if you looked at Venus. The surface is entirely obscured by thick clouds of sulfuric acid. If you looked through the clouds to the surface using radar, you would see the image on the right. Lots of volcanic features would be visible, and there would be relatively few craters compared to the Moon and Mercury.

Figure 15.2: Venus (http://www.nasa.gov, Magellan)

Venus' similarities to Earth quickly end when you get a good look at the planet. Figure 15.3 compares the Venusian atmosphere with the Earth's atmosphere. The Earth's atmosphere is composed primarily of nitrogen and oxygen, while Venus' atmosphere is composed primarily of carbon dioxide. Venus' atmosphere is nearly 100 times thicker than the Earth's atmosphere and is also very dry. Last, the Earth's average temperature at the surface is 60 degrees Fahrenheit; Venus' is 870 degrees Fahrenheit! This is hot enough to melt lead, and makes Venus the hottest planet...even hotter than Mercury!

	Earth	Venus
Composition	Mostly nitrogen and oxygen	Mostly carbon dioxide
Atmospheric pressure	1 atm at surface	90 atm at surface
Water	Wet (1-4% at surface)	Dry (0.002% at surface)
Surface Temperature	Average temperature of ~60 degrees Fahrenheit	Average temperature of ~870 degrees Fahrenheit

Figure 15.3: Comparison of Venus' and Earth's Atmosphere

To understand why Venus is so hot, we need to understand the greenhouse effect, and to understand that, we need to understand a little bit about what light is. Light is an electromagnetic wave. This means that it is a variation in the electric and magnetic fields that travels through space. However, the visible light that you see is only one kind of electromagnetic wave; there are others, like gamma rays and ultraviolet light. The difference between the types of electromagnetic waves is their wavelengths....in order from shortest to longest wavelengths, we have gamma rays, x-rays, ultraviolet rays, visible light, infrared waves, and radio waves (Figure 15.4). The frequency and energy of these waves is inversely proportional to the wavelength, so waves with a longer wavelength have a lower frequency and a lower energy.

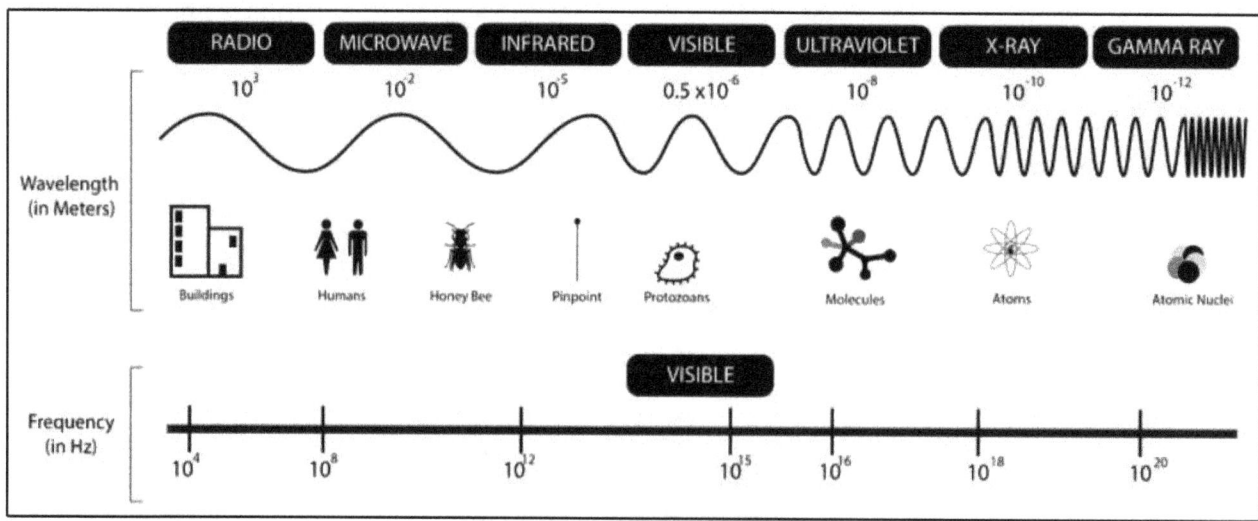

Figure 15.4: The Electromagnetic Spectrum
(https://commons.wikimedia.org/wiki/File:BW_EM_spectrum.png)

All objects absorb and emit electromagnetic radiation, however, they can absorb and emit different wavelengths. For example, your body absorbs visible light but emits infrared light (note that another word for infrared light is heat energy).

Planets do the same thing. Incoming energy from the Sun, which is mostly in the form of visible light, is absorbed by the surface and re-emitted at longer wavelengths, as infrared light. In the absence of the greenhouse effect, this infrared light escapes to space.

Now lets imagine we have a greenhouse as an analogy to how the greenhouse effect works. Glass is transparent to visible light (this is why you can see through a window), so the incoming solar energy is unaffected. However, glass absorbs infrared light (this is why heat doesn't leak out through the glass in windows in the winter), so the outgoing infrared light is trapped. This makes the greenhouse warm.

A greenhouse gas molecule is a gas molecule that behaves like glass: it is transparent to visible light, but absorbs infrared light. Carbon dioxide is an example of a greenhouse gas molecule. So the incoming solar energy is unaffected by the greenhouse gases, but the outgoing infrared light is trapped (Figure 15.5). It is not trapped forever; it will eventually make its way out. However, the more greenhouse gases there are, the longer the infrared light will be trapped, and the hotter the planet will be. Thus Venus is so hot because it has so much carbon dioxide. The reason it has so much carbon dioxide is because it was initially slightly too close to the Sun for liquid water. The Earth, which was slightly further away, had liquid water early on. This water absorbed the carbon dioxide and prevented a runaway greenhouse effect. Since Venus had no liquid water, the carbon dioxide remained, making the planet very hot.

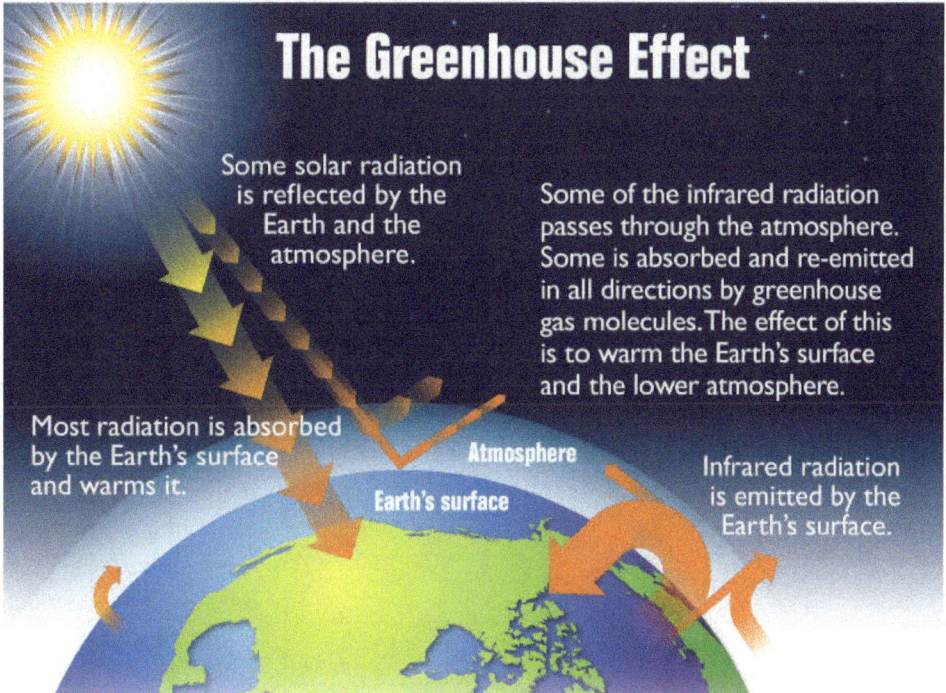

Figure 15.5: The Greenhouse Effect (Environmental Protection Agency)

Let's move to discussing the surface of Venus. Figure 15.6 shows a topographic map of Venus – blacks and dark grays are low elevations and light grays and whites are high elevations. Note that Venus has both smooth, low lying lava plains and highly deformed highlands (called Tesserae). Remember that when we discussed the Moon and Mercury, both planets had a range of ages on their surfaces; older, cratered areas and younger, less cratered areas. The same is not true for Venus! Anywhere you look on Venus, you will see roughly the same amount of craters. This means the entire surface of Venus is the same age. This age is very young (relative to the age of the solar system), only 500 million years old.

To illustrate how young this is, Figure 15.7 shows the ages of different planetary surfaces. The surfaces of Mercury, the Moon, and Mars are all older than 2 billion years. Only Earth and Venus have younger surfaces. However, even the Earth shows a range of ages. Only Venus shows a surface that is all pretty much all the same age! This is intriguing, and difficult for scientists to explain. It means that 500 million years ago, the entire surface of Venus was molten and cooled all at once. We can't know what happened before then (because there is no record preserved). However, some scientists hypothesize that Venus undergoes periodic catastrophic resurfacing, where the surface melts all at once. Other scientists hypothesize that Venus may have had

Figure 15.6: Topography of Venus
(http://www.nasa.gov, Magellan)

plate tectonics until a single resurfacing event. We just don't know.

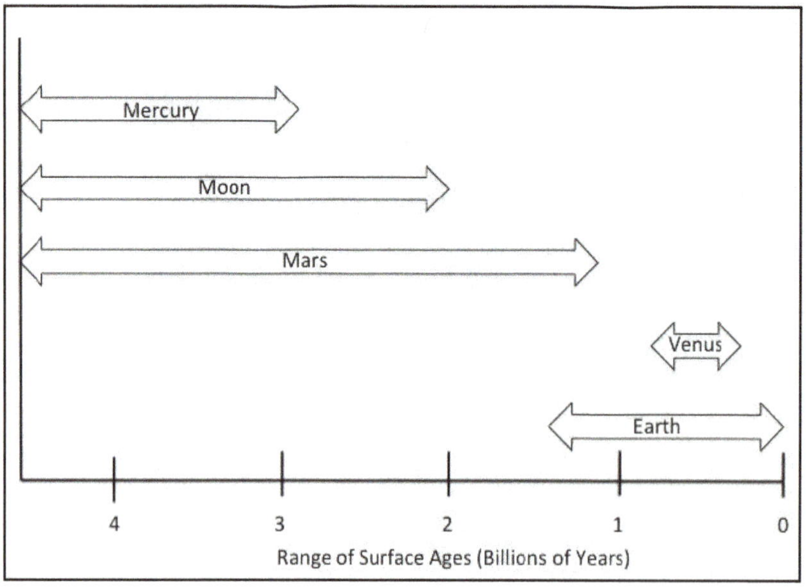

Because Venus is a similar size to the Earth and has active volcanic features, we can infer that it has a similar internal structure, with a molten outer core. However, Venus has no magnetic field, which is confusing. It means that either the outer core is solid (which doesn't make sense) or that the outer core is liquid but doesn't generate a magnetic field. Scientists have suggested that Venus spins too slowly for the liquid in the outer core to create a

Figure 15.7: Planet Surface Ages

magnetic field. This seems like a reasonable solution; however, it is unclear yet how the details of this would work.

We can sum this up into a possible history of Venus. Venus starts by outgassing a thick, carbon dioxide rich atmosphere. Because Venus is slightly too close to the Sun, it is slightly too warm for liquid water. Thus the atmosphere stays around and triggers a runaway greenhouse effect. We have no record of the following 4 billion years, because the entire surface is 500 million years old. Perhaps the planet undergoes period planet-wide overturning of the crust. Perhaps Venus behaved like the Earth (with plate tectonics) until a single recent resurfacing event. Other open questions are why are there no plate tectonics? And how did the tessera form? Venus is clearly a planet with a lot of discoveries left for scientists to figure out!

Questions You Should Be Able to Answer
1. How does Venus' mass, radius, and density compare to the Earth's?
2. How does the atmosphere of Venus differ from the atmosphere of the Earth?
3. Why is Venus so hot? In answering this question, be sure to explain the greenhouse effect.
4. Why is Venus' atmosphere so much thicker than the Earth's?
5. Describe the surface of Venus – how is it similar to and different than the surface of the moon and Mercury?
6. What reasons are there to think that Venus has a molten outer core? What reasons are there to think that Venus' interior is entirely solid?
7. What are some of the unknown questions concerning the history of Venus?

Suggestions for Further Reading
1. http://nineplanets.org/venus.html (Nine Planets Venus page)
2. http://nssdc.gsfc.nasa.gov/planetary/planets/venuspage.html (NASA Venus page)

Chapter 16: Mars

In this chapter, you will learn about Mars. We will talk about the exploration, atmosphere, surface, and history of Mars.

There have been many many missions to Mars, including flybs, orbiters, landers, and rovers (Figure 16.1). Note that about 1/3 of these mission actually failed, for various reasons. This highlights the fact that space missions are not particularly easy!

Missions to Mars

- Mariner 4, 6, 7, and 9 (1964, 1969, 1971): flybys and orbiter (9)
- Mars 2, 3, 5, and 6 (Soviet Union; 1971 and 1973): orbiters and landers
- Viking (1975): orbiters and landers
- Phobos (Soviet Union; 1988): orbiter and lander
- Mars Observer (MO) (1992): orbiter → contact lost three days before orbit insertion
- Mars Pathfinder (1996): lander and rover
- Mars Global Surveyor (MGS) (1996): orbiter
- Mars 96 (Soviet Union; 1996): orbiter, two landers, two impactors → crashed during takeoff
- Nozomi (Japan; 1998): orbiter → failed to go into orbit
- Mars Climate Orbiter (1998): orbiter → lost contact during orbit insertion due to unit conversion failure!!!!
- Deep Space Two (1999): two impactors → contact lost during flight to Mars
- Mars Polar Lander (1999): lander → crashed while attempting to land on Mars
- 2001 Mars Odyssey (2001): orbiter
- Mars Express (ESA; 2003): orbiter and lander (lander failed)
- Mars Exploration Rovers (MER) (2003): rovers (Spirit and Opportunity)
- Mars Reconnaissance Orbiter (2005): orbiter
- Phoenix (2007): lander
- Mars Science Laboratory (2011): rover (Curiosity) currently operating on Mars

Figure 16.1: Missions to Mars

Mars is smaller in size than the Earth; its mass is about 11% the Earth's mass and its radius is about 53% the Earth's radius. However, Mars' density is very similar to the Earth's density. The pictures in Figure 16.2 show you what you would see if you looked at Mars on different days. On the left is a clear day - you can see through the atmosphere to the features on the surface and the ice caps on the poles. The photo on the right was taken during a massive dust storm; wind picked up the dust on the surface and suspended it in the atmosphere, obscuring the view.

June 26, 2001   September 4, 2001
Hubble Space Telescope • WFPC2
NASA, J. Bell (Cornell), M. Wolff (SSI), and the Hubble Heritage Team (STScI/AURA) • STScI-PRC01-31

Figure 16.2: Mars (http://www.nasa.gov, Hubble)

Figure 16.3 compares Mars' atmosphere with the Earth's and Venus' atmospheres. Like Venus' atmosphere, Mars' atmosphere is composed primarily of carbon dioxide. However, while Venus' atmosphere is about 100 times thicker than the Earth's atmosphere, Mars' atmosphere is about 100 times thinner than the Earth's atmosphere. Mars' atmosphere is drier than the Earth's atmosphere, but wetter than Venus' atmosphere. Because Mars' atmosphere is so thin, the greenhouse effect is small, and the planet is very cold - the average temperature is -80 degrees Fahrenheit, much too cold for liquid water to exist. A cup of liquid placed on the surface would quickly freeze and then, because pressures are very low, would sublimate (turn into a gas) in the atmosphere.

	Earth	Venus	Mars
Composition	Mostly nitrogen and oxygen	Mostly carbon dioxide	Mostly carbon dioxide
Atmospheric pressure	1 atm at surface	90 atm at surface	0.006 - 0.01 atm at surface
Water	Wet (1-4% at surface)	Dry (0.002% at surface)	Dry (0.03% at surface)
Surface Temperature	Average temperature of ~60 degrees Fahrenheit	Average temperature of ~870 degrees Fahrenheit	Average temperature of ~-80 degrees Fahrenheit

Figure 16.3: Comparison of Earth's, Venus', and Mars' Atmospheres

Mars atmosphere was probably not always so thin and cold. It is estimated that it was 10-100 times thicker in the past. The gas that used to be in the atmosphere is now stored in the polar caps (which are composed of carbon dioxide ice with water ice underneath), stored as ice under the surface, stored as oxygen locked in rust (this is what makes the planet red), and lost to space.

Figure 16.4 shows a topographic map of Mars; darker gray is low elevation and lighter gary is high elevation. The planet appears divided roughly in half. In the northern area, we have smooth, young plains with few craters. This area is as smooth as the bottom of the Earth's oceans. In the southern area, we have the cratered highlands. There are several shield volcanoes dotting the surface, and a canyon, called Vallas Marineris runs along the equator. This canyon is the largest canyon in the solar system; at 2,500 miles long it would stretch from San Francisco to Boston.

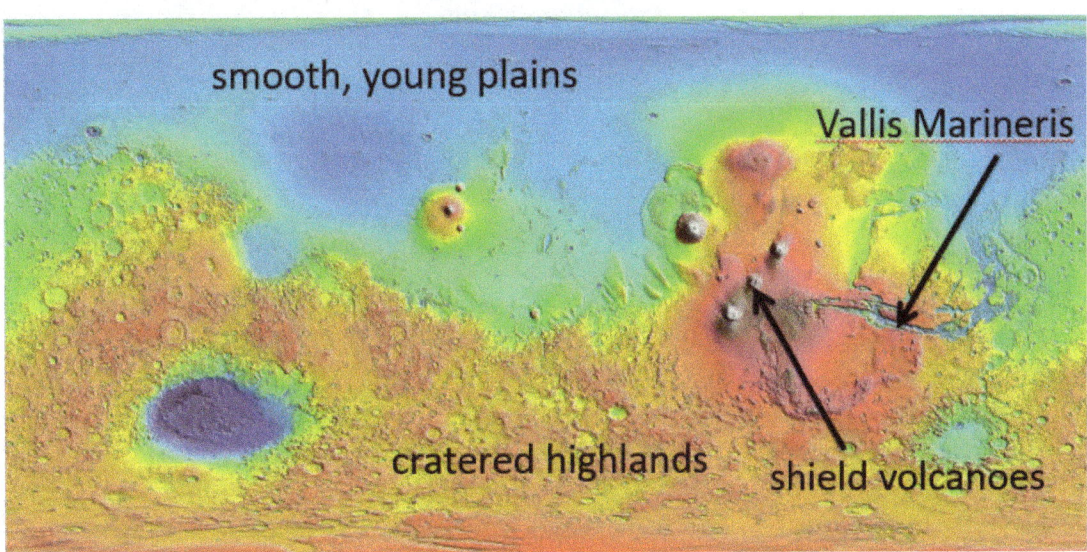

Figure 16.4: Topography of Mars (http://www.nasa.gov, Mars Global Surveyor)

There is abundant evidence that water once flowed over the Martian surface in the form of water carved and deposited features. We can find examples of outflow channels, stream valleys (Figure 16.5 left), and deltas (Figure 16.5 right) all over the surface of Mars.

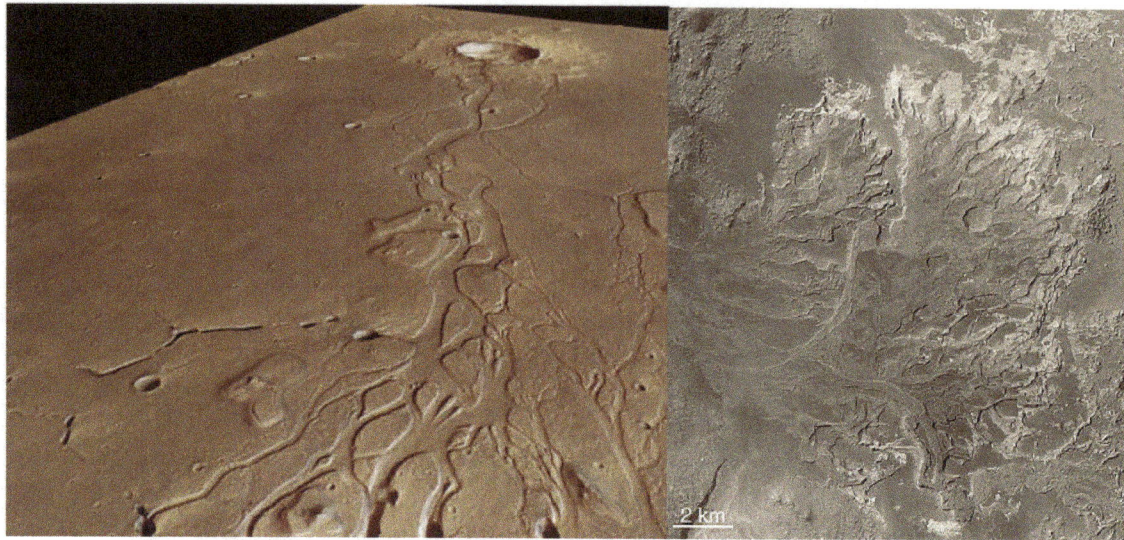

Figure 16.5: Stream Valley (left) and Delta (right) on Mars (ESA, Mars Express)

Additionally, there are rocks and minerals on the Martian surface that only form in the presence of liquid water. These include hematite spherules (nicknamed "blueberries"), sulfate minerals like jarosite, and sedimentary rocks (Figure 16.7).

If we look at the pattern of stream valleys, all of them appear to flow in the same general direction; downhill, towards the northern hemisphere, where they end. This suggests that the northern hemisphere may have once been filled with an ocean of liquid water. This would also explain the extreme smoothness of the northern plains.

Temperatures on Mars are currently too cold for liquid water. The water that once flowed over the surface in now stored in the polar ice caps, in glaciers, and underneath the surface.

Mars has a similar density to the Earth; thus its core/mantle ratio is similar. However, Mars is about half

Figure 16.6: Sedimentary Rocks on Mars (http://www.nasa.gov, MER Opportunity)

the radius of the Earth. Thus the interior of Mars has cooled and is no longer partially molten. Because of this, there is no magnetic field and no active volcanoes. However, there is evidence that there once was a magnetic field and active volcanoes (Figure 16.7).

Figure 16.7: Inactive Volcano (Olympus Mons) on Mars, and the Largest Volcano in the Solar System (http://www.nasa.gov)

We can put this together into a history of Mars. Scientists divide Martian history into three distinct periods. The early period, from formation (4.6 billion years ago) to 3.5 billion years ago, is known as the Noachian period. During this period, the Martian atmosphere was thicker. Because it was thicker, there was a stronger greenhouse effect and the planet was warmer. This allowed for liquid water on the surface. During this time the interior was still hot, allowing for abundant volcanic activity and a magnetic field. In other words, during the Noachian period, Mars was a lot like the Earth. The Southern hemisphere of Mars dates from this period.

The next period, from 3.5 to 1.8 billion years ago, is known as the Hesperian period. During this period, the interior cooled. As a result volcanic activity slowly decreased, there was less outgassing, and the magnetic field died. This had a profound effect on the atmosphere because the magnetic field protects the atmosphere. The sun continually gives of streams of charged particles - this is called the solar wind. If a planet has a magnetic field, it will deflect the solar wind safely around it. But without a magnetic field, the solar wind will continually bombard the atmosphere, leading to erosion as atmospheric particles are pushed into space. When Mars lost its magnetic field, its atmosphere began to thin from this process and was no longer replenished, since there was no more outgassing. As the atmosphere thinned, it cooled, and eventually temperatures were too cold for liquid water and any liquid water froze. The Northern hemisphere dates from this period and the next period.

The most recent period of Martian history, from 1.8 billion years ago to present, is known as the Amazonian period. During this period, the atmosphere has been thin and cold. The interior has been cooled, leading to no volcanism and no magnetic field. At this point, Mars is almost dead, with the only active processes being wind erosion and meteorite impacts.

Because Mars was a lot like the Earth at the same time that life was forming on the Earth, many scientists wonder if life ever developed on Mars. The most recent Mars Rover, Curiosity, focused onn this question. Curiosity found evidence of hospitable environments on Mars – lakes and streams that were not too salty and not to acidic or basic – just right for life. However, it's important to realize that if life did develop on Mars, it probably only reached the form of bacteria before the planet died (as that is how far along life on Earth got in that time frame), and fossils from bacteria are very difficult to identify, even on the Earth!

Questions You Should Be Able to Answer
1. How does Mars' mass, radius, and density compare to the Earth's?
2. How does the atmosphere of Mars differ from the atmosphere of the Earth and Venus?
3. Give some of the evidence showing that Mars may have had liquid water flowing over its surface in the past.

4. Give some of the evidence showing that Mars may currently have large reservoirs of ice under its surface.
5. Describe the history of Mars. Be sure to indicate the three major periods and give the major events that happened in those eras.
6. What is the name of the current rover exploring the surface of Mars?

Suggestions for Further Reading
1. http://nineplanets.org/mars.html (Nine Planets Mars page)
2. http://nssdc.gsfc.nasa.gov/planetary/planets/marspage.html (NASA Mars page)
3. http://mars.jpl.nasa.gov/ (JPL Mars page)
4. http://www.newyorker.com/magazine/2013/04/22/the-martian-chroniclers (excellent article about Curiosity from the New Yorker)

Chapter 17: Jovian Planets

In this chapter, you will learn about the Jovian planets. We will talk about the exploration, atmospheres, and interiors of these planets.

There have been relatively few missions to the outer solar system: Pioneer 10 and 11 and Voyager 1 and 2 performed flybys of Jupiter Saturn, Neptune, and their moons. Galileo orbited Jupiter, and Cassini orbited Saturn.

Missions to Jupiter, Saturn, Uranus, and Neptune

- Pioneer 10 and 11 (1973 and 1974): flybys of Jupiter
- Voyager 1 and 2 (1977-present day): flybys of Jupiter, Saturn, Neptune, and "beyond"
- Galileo (1989-1995): orbiter and probe to Jupiter
- Cassini (NASA/ESA joint mission; 1997-present day): orbiter to Saturn and probe to Titan

Figure 17.1: Missions to the Outer Solar System

Jupiter is much larger in size than the Earth; its mass is 318 times the Earth's mass and its radius is about 11 times the Earth's radius. Jupiter's density is much smaller than the Earth's density. This is because it is a Jovian planet, and is made mostly of gaseous elements (hydrogen and helium) and ice. When you look at Jupiter, you see clouds bands in a turbulent, churning atmosphere (Figure 17.2). To give you some size perspective, Figure 17.3 shows the Earth and Jupiter to scale. The Earth is about as big as the large swirling dot, which is known as Jupiter's "Great Red Spot."

Jupiter's atmosphere is composed primarily of hydrogen and helium (roughly 90% hydrogen and 10% helium with small amounts of other compounds). The different colors you see correspond to different cloud layers. Ammonia clouds are yellowish orange in color, ammonium hydrosulfide clouds are orangish red in color, and water vapor clouds are white in color. The banded pattern of the clouds is known as the belt-zone circulation and is created by the rapid rotation of the winds; Jupiter rotates once every 10 hours. The great red spot is a long-lived hurricane like storm which has been visible since Galileo first observed it 400 years ago.

Figure 17.2: Jupiter
(http://www.nasa.gov, Voyager)

Shown in Figure 17.4 is a cross-section through Jupiter's atmosphere. Altitude is shown on the y-axis, and temperature is shown on the x-axis (ignore the Uranus line for now). Note that since Jupiter has no solid surface, 0 altitude has no meaning and is randomly defined. Deep in the atmosphere, temperatures are hot enough that everything is a gas. As you go up, the temperature cools. When you hit 273 Kelvin, water ice condenses out into clouds. When you hit 200 Kelvin,

Figure 17.3: Jupiter (http://www.nasa.gov, Voyager) with the Earth to Scale

ammonium hydrosulfide ice condenses out into clouds. When you hit 130 Kelvin, ammonia ice condenses out into clouds. Above that there is a haze layer where the atmosphere is filled with dust particles, creating a hazy smog.

What about the interior of Jupiter? There are a few clues. The first is that Jupiter has a very strong magnetic field; it is 14 times larger than the Earth's field at the surface. Something in the interior must be generating this field. Second, the density of Jupiter constrains the rocky/metallic core to 1-10 Earth masses. This seems large, until you remember than the entire planet is 318 Earth masses. Third, Jupiter's radius at the equator is larger than its radius at the poles; it is slightly fat in the middle. This oblateness indicates that the interior is largely liquid. Fourth, we can use mathematic models to learn about Jupiter's interior.

Let's imagine that you could parachute through Jupiter's atmosphere. Jupiter's temperature and pressure are beyond what is known as the critical point of hydrogen. This means there is no distinction between the gaseous and liquid states. Thus, as you parachuted, the atmosphere around you would gradually get thicker and thicker. Eventually you would look around and realize that you are in a liquid; but there would have been no abrupt transition or splashdown. A quarter of the way to the center, the liquid hydrogen would transition into liquid metallic hydrogen. This means that the electrons in the hydrogen atoms are free to float from atom to atom, like in a metal. The liquid

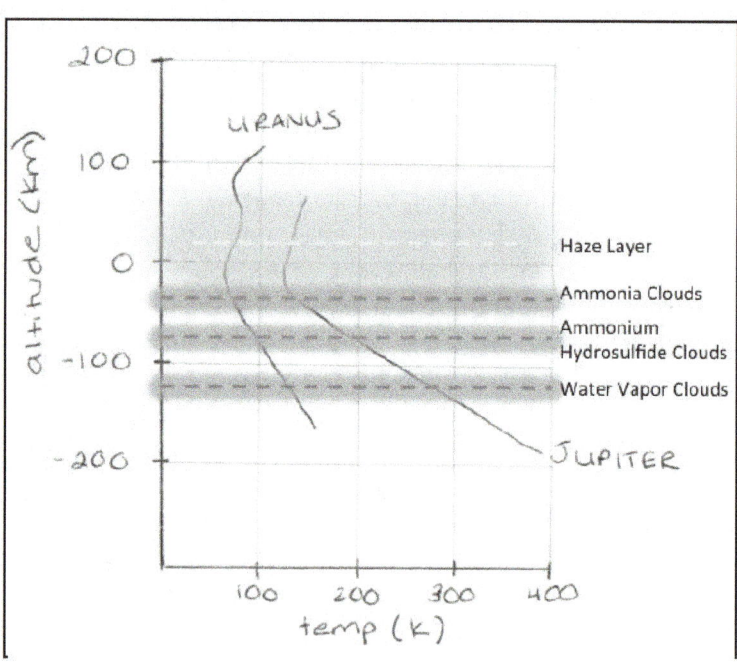

Figure 17.4: Cross-Section Through Jupiter's Atmosphere

metallic hydrogen thus acts exactly like a metal, and this generates a very strong magnetic field, which is about 20 times stronger than the Earth's. By the time you reach the center of Jupiter, the temperature would be 5 or 6 times hotter than the Sun!

Jupiter's structure is illustrated in Figure 17.5 diagram (which also has the structure of the Earth to scale for comparison). We have the cloud tops, the gaseous hydrogen atmosphere, the liquid hydrogen, the liquid metallic hydrogen, and a rocky/metallic core at the center. Note that it is a misnomer to call Jupiter a gas giant - as you can see, it is mostly liquid!

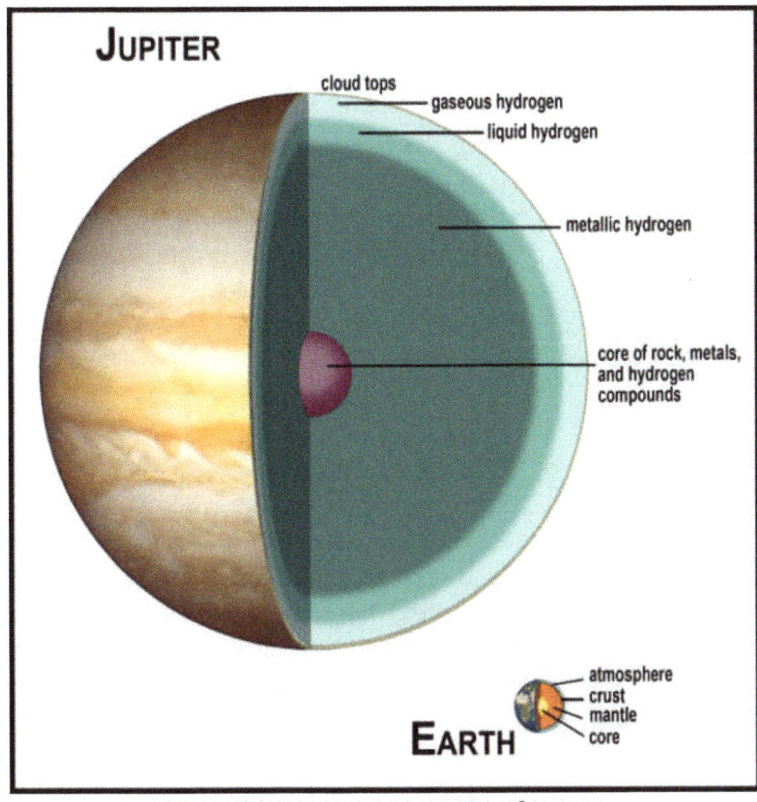

Figure 17.5: Interior Structure of Jupiter
(NASA Lunar and Planetary Institute)

Jupiter has rings, but they are difficult to see, and can only be seen when the planet is lit from behind. The rings are composed of rocky, microscopic particles.

Saturn is much larger in size than the Earth and is similar in size to Jupiter; its mass is 95 times the Earth's mass and its radius is about 9 times the Earth's radius (Figure 17.6).

While Saturn is similar to Jupiter, there are a few significant differences. The first is that Saturn is somewhat smaller, about 1/3 the size of Jupiter. This leads to a weaker magnetic field, since there is not as much liquid metallic hydrogen. Second, Saturn's belt-zone circulation is not as well-defined; it looks "smoother" in images. This is because Saturn is smaller and colder; thus, the cloud layers form deeper down under the haze layer and are harder to see. Third, Saturn's rings are much more prominent than Jupiter's.

Saturn's rings are mostly composed of particles of ice that range in size from dust to small moons. There are dramatic gaps in the rings. To understand why Saturn's rings are so different from Jupiter's rings, we need to understand where the rings come from. Scientists initially thought that the rings were leftovers from solar system formation. However, this would not explain why Jupiter has only faint rings. Furthermore, computer simulations show that rings should clear out in 10 to 100 million years, much shorter than the age of the solar system.

We now think that rings are created in one of two ways. Either a moon gets too close to a planet, and the planet's gravity rips it apart...or, two moons collide and rip each other apart. The resulting debris becomes a ring. Thus Saturn has prominent rings because one of these two events must have happened relatively recently.

Figure 17.6: Saturn
(http://www.nasa.gov, Voyager)

A final feature that we see on Jupiter and Saturn is auroras. Auroras are patterns of light that you can see at the north and south poles of a planet; we have them on Earth (Figure 17.7). Auroras are created when solar wind from the Sun is funneled to the poles by the planet's magnetic field. As the solar wind particles come in at the poles, they collide with molecules in the atmosphere. The collisions release energy in the form of light.

Uranus and Neptune are very similar to eachother. Both are larger in size than the Earth, but smaller than Jupiter; Uranus' mass is 14 times the Earth's mass, but only 5% of Jupiter's mass. Its radius is about 4 times the Earth's radius. When you look at Uranus, you see a featureless blue ball. Neptune's mass is 17 times the Earth's mass, and its radius is about 4 times the Earth's radius. When you look at Neptune, you see a blue ball with some faint banding (Figure 17.8).

Figure 17.7: Auroras at the Earth's North Pole
(http://image.gsfc.nasa.gov/)

Uranus and Neptune are similar to Jupiter and Saturn with a few notable differences. The first is color. Jupiter and Saturn are orangish-red, while Uranus and Neptune are blue. This is because different cloud layers are visible. When you look at Jupiter and Saturn, what you are seeing are clouds of ammonia, ammonium hydrosulfide, and water. When you look at Uranus and Neptune, what you are seeing is clouds of methane. Second, Uranus and Neptune are relatively featureless. This is not because there are no wind bands; there are. However, because there is only one cloud layer visible, you can't see the wind bands. Third, Uranus and Neptune have very weak magnetic fields (weaker even than the Earth's). Both have rings, but they are faint (like Jupiter's).

Figure 17.8: Uranus (left) and Neptune (right)
(http://www.nasa.gov, Voyager)

The reason that Uranus and Neptune have weak magnetic fields is because they are smaller. Since they are smaller, their interiors never reach the high pressures and temperatures required for liquid metallic hydrogen. Their magnetic fields are instead generated by dissolved ions moving around in liquid water near the surface. This is a very weak source of a magnetic field. Figure 17.9 compares the interiors of all of the Jovian planets. Uranus has a thick envelope of gaseous hydrogen and helium, then a layer of liquid ice and rock, and heavy element core at the center.

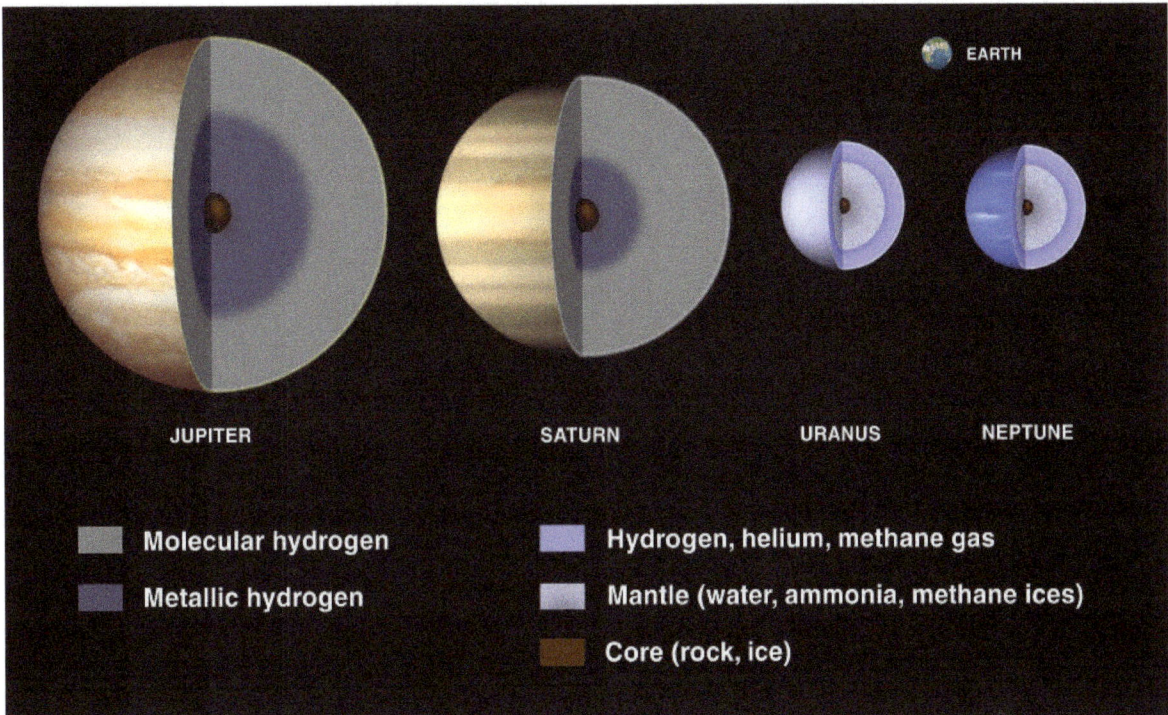

Figure 17.9: Interiors of the Jovian Planets (http://www.nasa.gov)

Questions You Should Be Able to Answer
1. How does Jupiter's mass, radius, and density compare to the Earth's?
2. How does the atmosphere of Jupiter differ from the atmosphere of the Earth?
3. Describe the interior of Jupiter and how scientists know anything about it.
4. How does Saturn's mass, radius, and density compare to the Earth's and to Jupiter's?
5. Give some similarities and differences between Jupiter and Saturn.
6. Why is Saturn's magnetic field weaker than Jupiter's magnetic field?
7. Why is wind banding less prominent on Saturn than on Jupiter?
8. Why are Saturn's rings so much more prominent than Jupiter's rings?
9. Explain what creates wind bands, auroras, rings, and spots (like the Great Red Spot of Jupiter).
10. How does Uranus' and Neptune's mass, radius, and density compare to the Earth's and to Jupiter's?
11. Why are Uranus and Neptune blue in color?
12. Why are Uranus' and Neptune's atmospheres relatively featureless (compared to Saturn's and Jupiter's atmosphere's)?
13. Why are the magnetic fields of Uranus and Neptune so weak?

Suggestions for Further Reading
1. http://nineplanets.org/jupiter.html (Nine Planets Jupiter page)

2. http://nineplanets.org/saturn.html (Nine Planets Saturn page)
3. http://nineplanets.org/uranus.html (Nine Planets Uranus page)
4. http://nineplanets.org/neptune.html (Nine Planets Neptune page)
5. http://nssdc.gsfc.nasa.gov/planetary/planets/jupiterpage.html (NASA Jupiter page)
6. http://nssdc.gsfc.nasa.gov/planetary/planets/saturnpage.html (NASA Saturn page)
7. http://nssdc.gsfc.nasa.gov/planetary/planets/uranuspage.html (NASA Uranus page)
8. http://nssdc.gsfc.nasa.gov/planetary/planets/neptunepage.html (NASA Neptune page)

Chapter 18: Outer Solar System Satellites

In this chapter, you will learn about the moons of the outer solar system. We will talk specifically about the large moons of Jupiter and Saturn.

Figure 18.1 lists the number of moons each planet has, as well as some of their names. The terrestrial planets have few moons; Mercury and Venus have none, the Earth has one, and Mars has two. However, the Jovian planets have many moons; Jupiter has 62, Saturn has 33, Uranus has 27, and Neptune has 13. Notice the interesting naming conventions of these moons. Most are named after characters from Greek and Roman mythology: Jupiter's moons are named after Jupiter's/Zeus' mortal lovers, Saturn's moons are named after the Titans and characters related to the Titans, and Neptune's moons are named after characters related to Neptune/Poseidon. The oddball out is Uranus, whose moons are named after Shakespeare characters!

PLANET	NUMBER OF MOONS	SOME OF THE NAMES
Mercury	0	
Venus	0	
Earth	1	Moon
Mars	2	Phobos, Deimos
Jupiter	62	Io, Europa, Ganymede, Callisto,, Amalthea, Himalia, Elara, Pasiphae, Sinope, Lysithea
Saturn	33	Titan, Rhea, Iapetus, Dione, Tethys, Prometheus, Pandora, Phoebe, Enceladus, Mimas
Uranus	27	Cordelia, Ophelia, Bianca, Cressida, Desdemona, Juliet, Portia, Rosalind, Belinda, Puck
Neptune	13	Triton, Nereid, Naiad, Thalassa, Despina, Galatea, Larissa, Proteus

Figure 18.1: Moons In our Solar System

Selected moons are shown to scale in Figure 18.2. The largest moons are about the size of the Earth's moon. This may seem surprising: you might think that since Jupiter is bigger, that it would have larger moons. The reason behind this is probably the formation mechanism. The moons of the Jovian planets are thought to have formed via binary accretion and the moons of Mars are thought to be captured asteroids. Thus the Earth's moon is so big relative to its planet because it is the only one formed from a giant impact. Jupiter has four large moons, Saturn has one large moon, and the rest of the moons are fairly small. We will talk specifically about each of the large moons.

However, before we do, we should note that the moons are shaped by the same things the Earth and other Terrestrial planets are. Things like impact cratering, volcanism, wind, plate tectonics, and water.

Figure 18.2: Selected Moons of the Solar System to Scale
(https://commons.wikimedia.org/wiki/File:Moons_of_solar_system_v7.jpg)

Additionally, there are two processes that operate on these moons that do not operate on the Terrestrial planets. The first is that strong tides can cause a hot, molten interior, even on a small moon. The second is that liquids other than water can occur on moons that are too cold for liquid water.

Figure 18.3 shows a diagram to help you understand how tides can cause a hot, molten interior. In the diagram, we have a moon orbiting around Jupiter (note that the diagram is not to scale). The pull of Jupiter's gravity on the moon is shown with the black arrows; a longer arrow represents a greater pull. Jupiter's gravity pulls the moon towards it. Because the side of the moon that is facing Jupiter is pulled more than the side of the moon facing away from Jupiter (this is because it is closer, and gravity is stronger when objects are closer), this stretches the moon out into an oval shape. Additionally, since the moon's orbit is elliptical, sometimes it is closer to Jupiter and sometimes it is further. When it is closer, it is more stretched out. When it is further, it is less stretched out. Thus as the moon orbits Jupiter, it repeatedly goes from being less to more stretched out; this is analogous to pulling a rubber band back and forth. This motion will generate friction that heats the planet (just like the rubber band gets hot). If the friction generated is enough, it can even melt the interior and cause volcanoes!

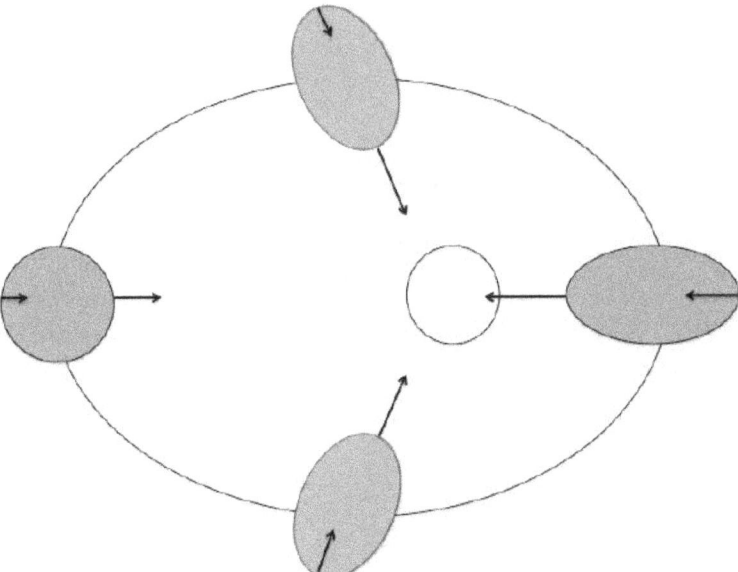

Figure 18.3: Tidal Heating

Titan is Saturn's large moon. When you look at Titan, you can't see through the atmosphere to the surface because it has a very thick nitrogen atmosphere. The surface is composed of frozen water ice and frozen methane. The methane on Titan acts like water on the Earth. It evaporates off the surface, forms clouds, and rains back down onto the surface where it pools in rivers and lakes. There are few craters on the surface because of this. We know that life on Earth needs liquid water to survive; however, many scientists wonder if liquid methane on Titan could substitute for liquid water. Thus Titan is a good place to look for evidence of life.

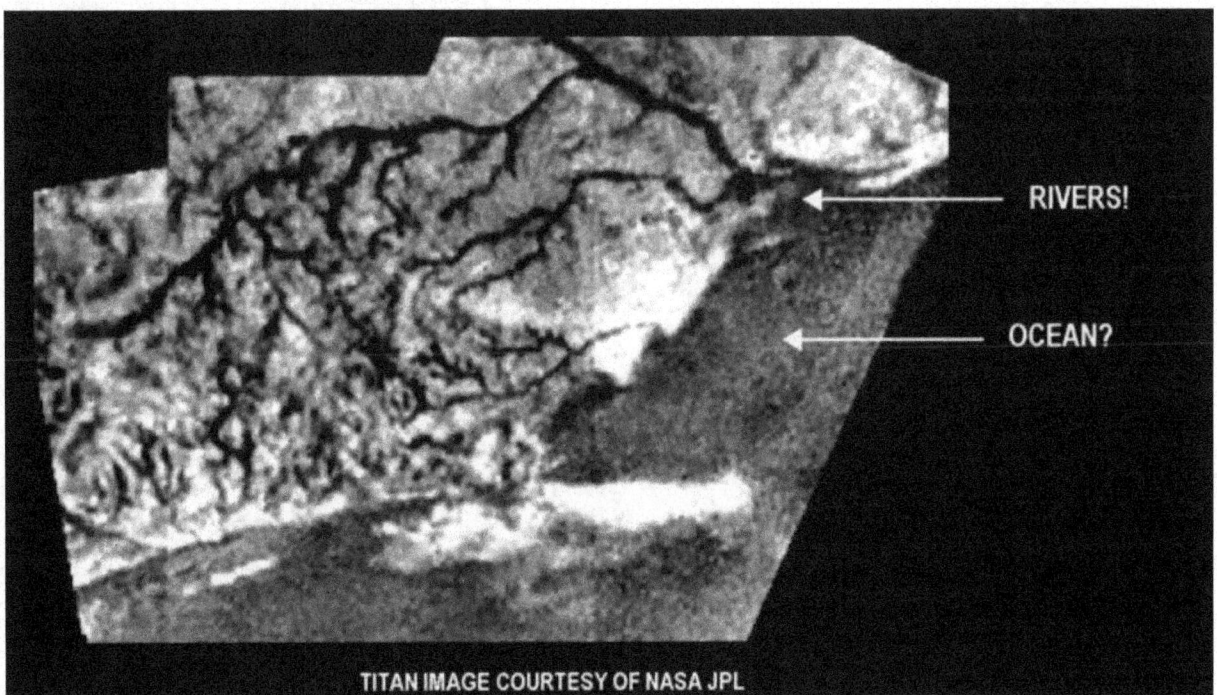

Figure 18.4: Surface of Titan Showing Rivers of Liquid Methane and a Possible Ocean/Lake
(http://www.nasa.gov, Huygens Probe)

Jupiter has four large moons. They are Io, Europa, Ganymede, and Callisto. Though they are all very similar in size (about the same size as the Earth's moon), we are going to see that they are very very different.

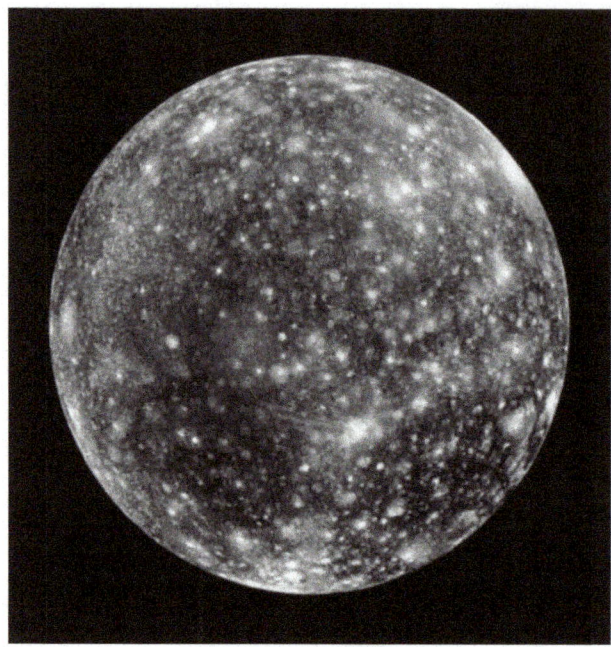

Figure 18.5: Callisto
(http://www.nasa.gov, Voyager)

Let's start with Callisto. Callisto has a density of 1800 kg/m^3. This indicates that it is made of ice and rock. When you look at Callisto, you see an old cratered surface. There are darker and lighter areas. The darker areas are dirty ice, and the lighter areas are where recent impacts have exposed cleaner ice that lies beneath. The longer ice sits on the surface of the moon, the dirtier it will get because the ice will sublimate away, leaving dirt behind.

Callisto's interior is uniform and undifferentiated. This means that the ice and rock is all mixed homogenously together; it is not divided into layers.

In contrast, the other large moons of Jupiter are differentiated (Figure 18.6). Both Europa and Ganymede have icy surfaces, followed by a layer of liquid water, followed by a rocky mantle, followed by a metallic core at the center. Io has no water (frozen or liquid) - just a rocky mantle and a metallic core. The reason for these differences has to do with the moons' distances to Jupiter. The closer a moon is to Jupiter, the more tidal heating it will experience. Callisto is the furthest from Jupiter, so tidal heating is insignificant. Thus Callisto never got hot enough to differentiate. Ganymede is second furthest, and then Europa. Both of these moons experienced enough tidal heating to cause them to differentiate and to melt the liquid water into subsurface oceans. Io is the closest to Jupiter and so experiences the most tidal heating. The tidal heating on Io is so strong that it caused Io to differentiate, boiled the liquid water away, and melted some of the rocks in the interior, giving Io a hot, molten interior.

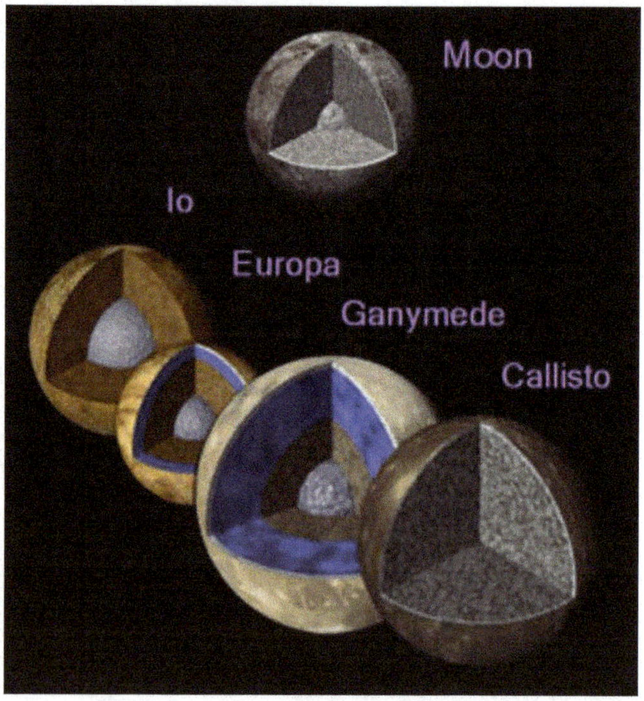

Ganymede has an icy surface composed of darker, older terrain, and brighter, younger terrain (Figure 18.7). There are cracks in the younger terrain where water from the ocean underneath welled up and flooded the surface. Salts have been detected on the surface,

Figure 18.6: Interior's of Jupiter's Moons
(http://www.amnh.org/)

indicating that the ocean is salty. The ocean is estimated to be 3 miles thick and 110 miles below the surface.

Europa is similar to Ganymede. However, there is no darker, older terrain. All of the surface is young (Figure 18.8); this is because Europa's ocean is closer to the surface than Ganymede's; it is estimated to be 100 miles thick and 10 miles below the surface. Thus upwelling through cracks occurs more often. Recent measurements have found that Europa intriguingly has a thin, oxygen atmosphere. It's origin is unknown, but it is intriguing because we know that it is life (particularly photosynthetic organisms) that are the source of Earth's oxygen.

Life exists at the bottom of the Earth's oceans, where no sunlight reaches; this life receives its energy from volcanic vents that release heat. Scientists hypothesize that similar vents should exist on the bottom of Europa's ocean, which makes Europa another good place to look for evidence of life.

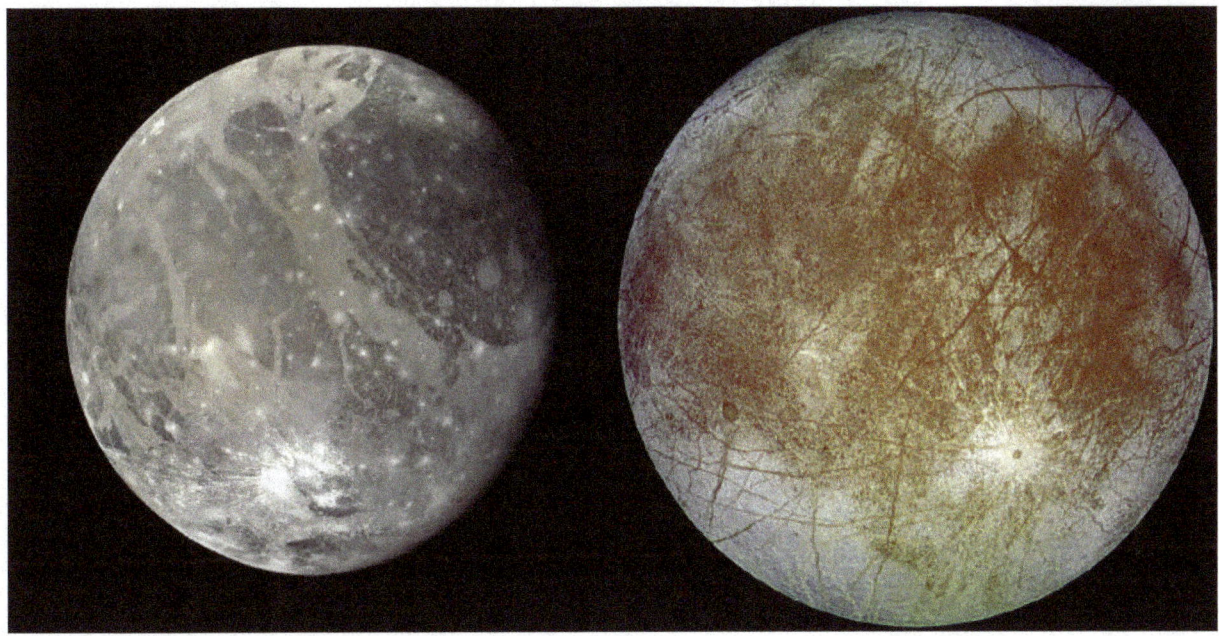

Figure 18.7: Ganymede
(http://www.nasa.gov, Voyager)

Figure 18.8: Europa
(http://ww.nasa.gov, Voyager)

Io is the driest and most geologically active body in the entire solar system. The tides on Io are so strong that the surface rises and falls by 330 feet each Io day (which is 1.8 Earth days). This generates so much molten rock in the interior that volcanoes on Io are ubiquitous. We have pictures of volcanoes erupting (there are two in the image in Figure 18.9), of lava flowing on the surface (Figure 18.10), and of the gases and dust released by these volcanoes. Eruptions on Io are so frequent that not a single impact crater has been found.

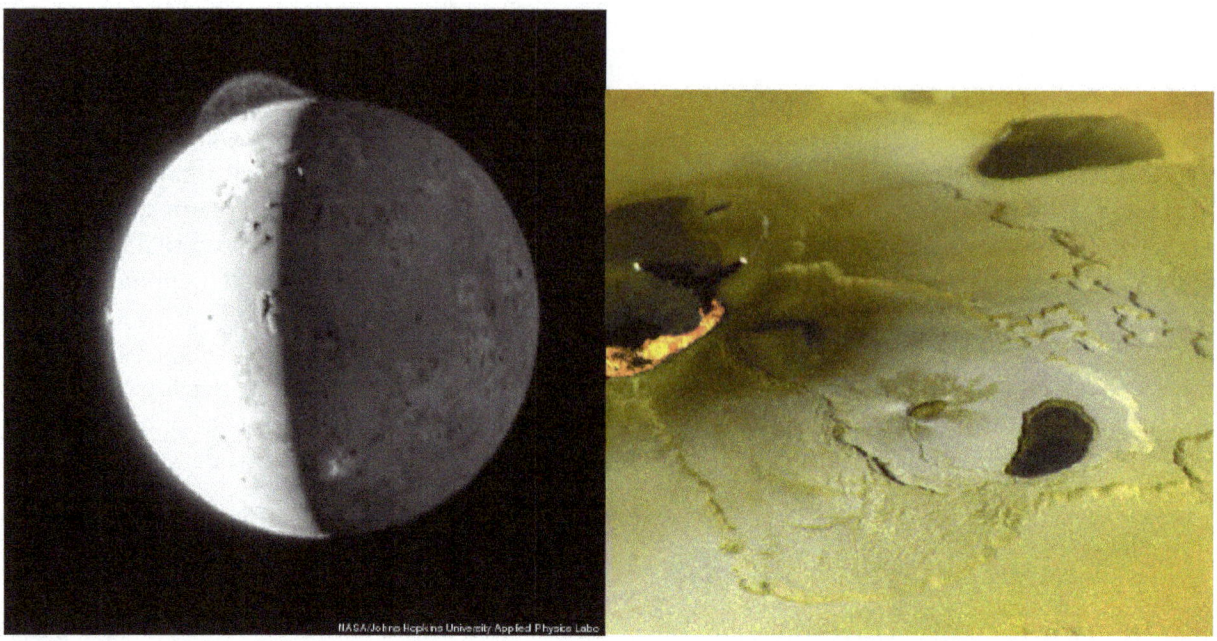

Figure 18.9: Io
(http://www.nasa.gov, New Horizons)

Figure 18.10: Lava on Io
(http://www.nasa.gov, Galileo)

Questions You Should Be Able to Answer
1. What geologic processes shape the moons of the outer solar system?
2. What is tidal heating?
3. Describe the characteristics of Titan.
4. How are measurements of a planet's moment of inertia useful in determining a planet's interior structure?
5. Describe the surfaces and interiors of Io, Europa, Ganymede, and Callisto. Explain what factors cause the extreme differences.

Suggestions for Further Reading
1. http://nineplanets.org/io.html (Nine Planets Io page)
2. http://nineplanets.org/europa.html (Nine Planets Europa page)
3. http://nineplanets.org/ganymede.html (Nine Planets Ganymede page)
4. http://nineplanets.org/callisto.html (Nine Planets Callisto page)
5. http://nineplanets.org/titan.html (Nine Planets Titan page)
6. http://www.astro.washington.edu/users/smith/Astro150/Tutorials/TidalHeat/ (Tidal Heating Explained)

Chapter 19: Detecting Exoplanets

In this chapter, you will learn about the methods that scientists use to find exoplanets (also known as extrasolar planets). We will talk specifically about the radial velocity and transit methods.

An exoplanet is a planet the orbits a star other than our Sun. This is a relatively new field of study. The first exoplanet was discovered in 1989, but we did not begin to find them in abundance until the turn of the millennium. As of June 6, 2018, there were 2950 confirmed exoplanets and 2337 unconfirmed candidates.

Finding exoplanets, particularly Earth-sized exoplanets, is not an easy task. Imagine a basketball on the Moon. Then imagine a grain of sand orbiting that basketball. Trying to find that grain of sand without leaving the Earth is analogous to trying to find an Earth-size exoplanet!

So how do scientists do it? The short answer is that they use telescopes. Both Earth-based telescopes and Space-based telescopes. Of the Space-based telescopes, Kepler is the most notable because it is the only telescope so far that was dedicated solely to finding exoplanets. Unfortunately, Kepler malfunctioned in 2013 and is no longer actively collecting data.

How do telescopes find exoplanets? There are several methods, including radial velocity, transit, microlensing, imaging, and pulsar timing. Very few exoplanets have been discovered by imaging, which is directly taking a picture of a planet. This is because most exoplanets are too small and dim to image. Thus, the other methods focus on watching a star and inferring that a planet is there by looking for certain changes in the light that the star is emitting. Most exoplanets have been discovered using two methods: radial velocity and transit. We will discuss these methods in detail.

To understand the radial velocity method, we need to understand the Doppler Effect. The Doppler Effect is the shift in wavelength observed when an object that is giving off light is moving towards you or away from you. First, imagine a star that is not moving. In this scenario, the light that is emitted by the star is exactly what you see. Second, imagine that a star is moving away from you; we say it has a positive radial velocity. In this scenario, the light that you see is longer in wavelength than the light that the star is emitting because it gets stretched out. We call this a redshift, because red light has longer wavelength than other visible light. Please note that this does NOT mean that the star looks red; just that the observed light from the star is at a longer wavelength than what the star is actually emitting. The faster the star is moving away from you, the bigger the redshift. Third, imagine that a star is moving towards you; we say it has a negative radial velocity. In this scenario, the light that you see is shorter in wavelength than the light that the star is emitting because it gets squeezed together. We call this a blueshift, because blue light has shorter wavelength than other visible light. The faster the star is moving towards you, the bigger the blueshift.

Note that the Doppler Effect only happens if the object that is emitting light is moving towards you or away from you. An object that is moving side to side, or perpendicular to your line of sight, will not show any shift.

An object's motion along your line of sight is known as its radial velocity (Figure 19.1). By convention, when an object is moving away from you, we say that it has a positive radial velocity. When an object is moving towards you, we say that it has a negative radial velocity. When an object is moving perpendicular to you, we that is has zero radial velocity. Since the amount of red or blueshift depends on an object's speed, looking at the Doppler Shift of a star gives you its radial velocity.

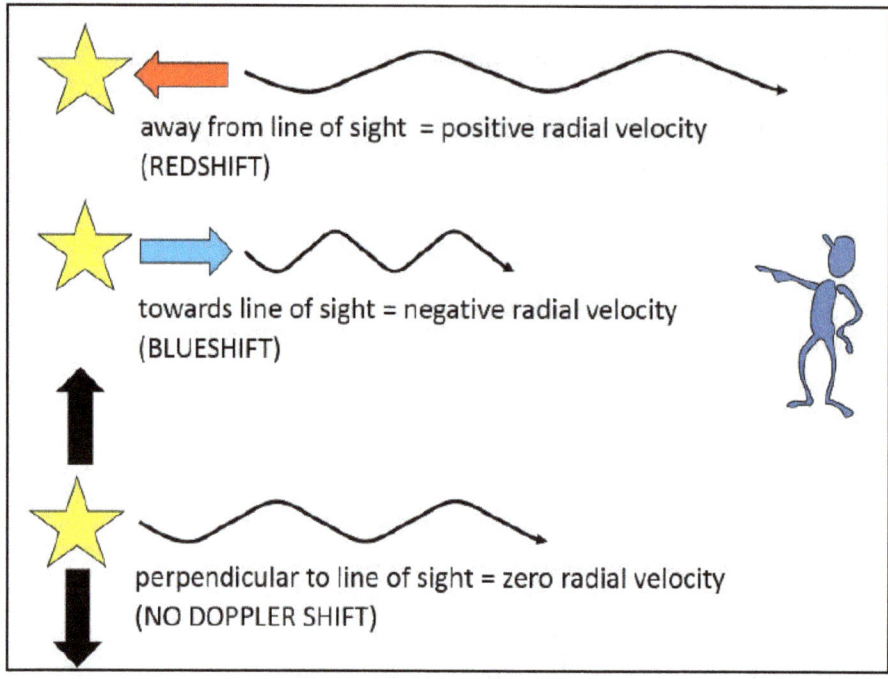

Figure 19.2: Doppler Effect

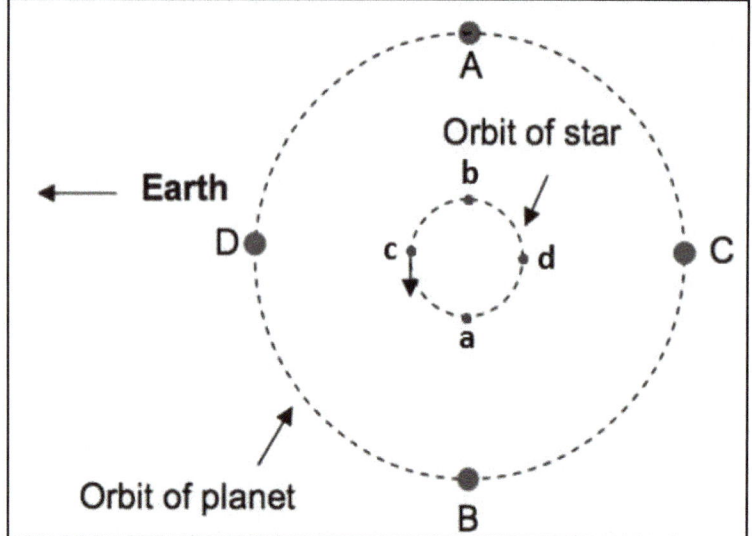

Figure 19.2: Exoplanet and Star Orbiting Center of Mass

How does this apply to exoplanets? We know that planets orbit around stars; but this statement is slightly inaccurate. In reality, what is actually happening is that both the planet and the star are orbiting around the common center of mass (Figure 19.2). Though the planet's orbit is larger than the star's orbit, both the star and the planet move around the center of mass in the same time. So for example, when the planet is at location "A" in Figure 19.2, the star is at location "a". This means that the star and the planet are always at opposite positions in their orbits. Now imagine you are on Earth (which is to the left in Figure 19.2) looking at an edge-on view of this solar system. You can't see the planet; it's too dim. But you can see the star. Note that sometimes the star is going towards you (at "b"), sometimes it is going away from you (at "a"), and sometimes its motion is perpendicular to your line of sight (at "c" and "d"). Thus as you watch the star, you will see its light shift from blueshifted to no shifted to redshifted to no shifted to blueshifted and so forth. Because the star would not do this if it didn't have a planet around it, you can infer that the planet is there. Using the amount of shift, you can calculate the star's exact radial velocity over time. From this information, you can infer the mass of the planet and the planet's distance from the star.

The star's radial velocity curve will look something like shown in Figure 19.e. The peaks correspond to the points when the star is moving at its maximum away from you. The dips corresponds to the points when the star is moving at its maximum towards you. The zero points correspond to when the star is moving neither towards nor away from you; it is moving perpendicular to your line of sight.

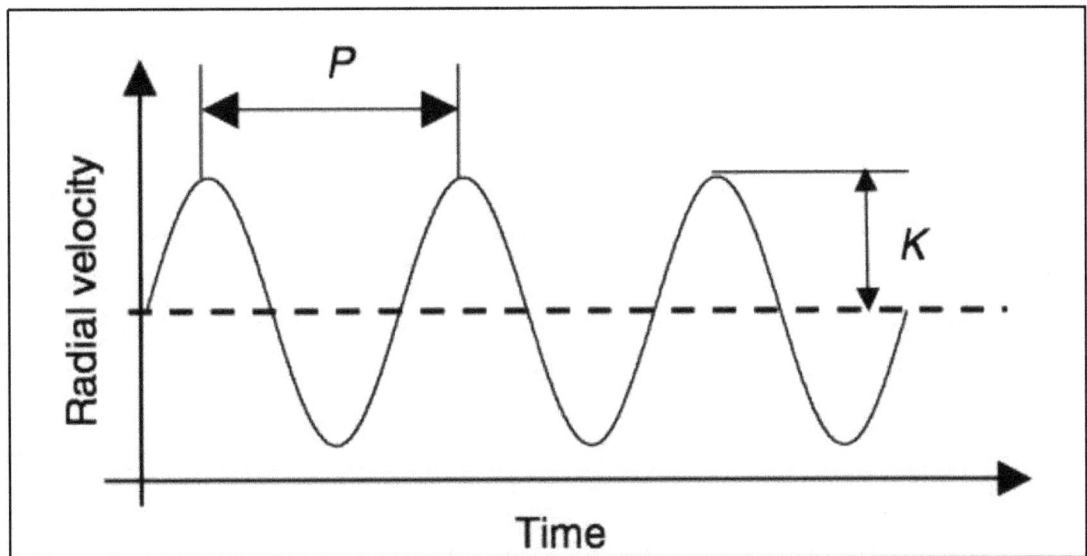

Figure 19.3: Radial Velocity Curve for A Star With an Exoplanet

The period of the radial velocity curve (labeled "P" in Figure 19.3) is the time from peak to peak. It represents how long it takes the star and the planet to complete one orbit. Remember from Kepler's 3rd law that a planet's period depends on its distance. Thus from the period, you can calculate how far the planet is from its star.

The amplitude (labeled "K" in Figure 19.3) is the height of the curve. It depends on the planet's mass, the star's mass, and the planet's distance. The bigger the planet, the bigger the amplitude. This is because the planet will have a greater pull on its star. The smaller the star, the bigger the amplitude. This is because big stars will have more resistance to motion. The closer the planet, the bigger the amplitude. This is because if the planet is closer it will have a greater pull. Thus bigger amplitudes correspond to large planets that are close to small stars.

Besides the radial velocity method, the other common method of exoplanet detection is the transit method. If you are viewing a solar system from an edge-on view, then a planet will occasionally move in front of its star...this is known as a transit. When it does so it will block a little bit of the light coming from the star, and the star's light will become slightly dimmer. This drop in light is typically 0.01%-1%. By detecting and tracking these dips we can infer the presence of a planet and calculate its orbital distance (planets that are further will show more time between dips and wider dips) and its radius (planets that are bigger will cause deeper dips because they block more light).

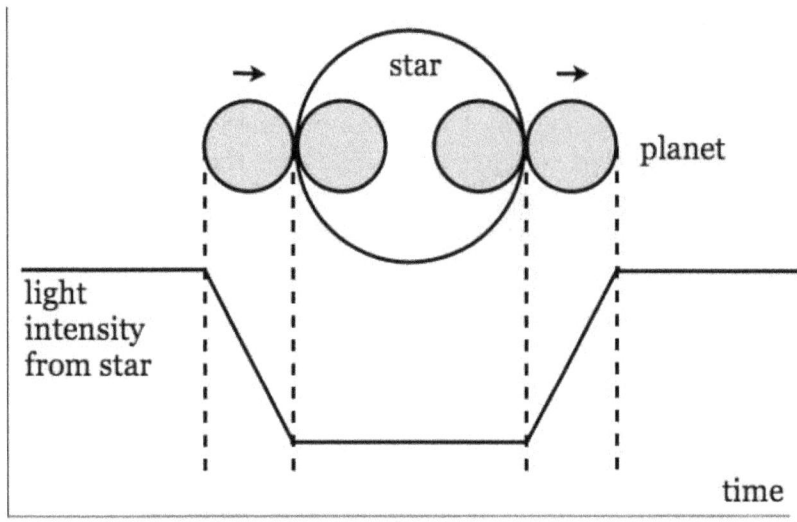

Figure 19.4: Exoplanet Transiting a Star (CAE Astronomy Workshop Materials)

This is the method of detection used by the Kepler space telescope. The Kepler space telescope searched for exoplanets within a cone of 3,000 light years away from the Earth. This is just a very small portion of galaxy. Note that all of the exoplanets that have been discovered are within are galaxy; other galaxies are much too far away for us to detect planets. If you are curious where in the sky this is, look outside at the night sky in the summer. There are three bright stars that form a triangle; this constellation is known as the summer triangle. Kepler's field of view is right at the base of the triangle.

The image shown in Figure 19.4 describes the transit method in more detail. Initially the light coming from the star shows a constant intensity. As the planet begins to pass in front of the star, the intensity dips. It reaches a minimum when the planet is completely in front of the star. When the planet begins to move beyond the star, the intensity will gradually increase back to its maximum.

The transit dip has several features, each which represent different aspects of the planet. The dip depth is determined by how much light is blocked. This is determined by the planet's radius; larger planets will block more light and cause a deeper dip. The width of the dip is determined by how long the transit takes. This is determined by the planet's orbital period; longer orbital periods will lead to a longer transit and cause a wider dip (because the planet is moving slower). The sides of the dip are sloped because it takes some time for the planet to move in front of and out from in front of the star. The bottom of the dip is flat because when the planet is completely in front of the star, it blocks a constant amount of light.

The time from the beginning of one dip to the beginning of the next dip is the time it takes the planet to complete one orbit - the orbital period (Figure 19.5). Remember from Kepler's 3rd Law that the orbital period is determined by the planet's distance from the Sun. Thus planets that are further will have a greater time between dips. Note that the time between dips is always longer than the duration of the dips themselves.

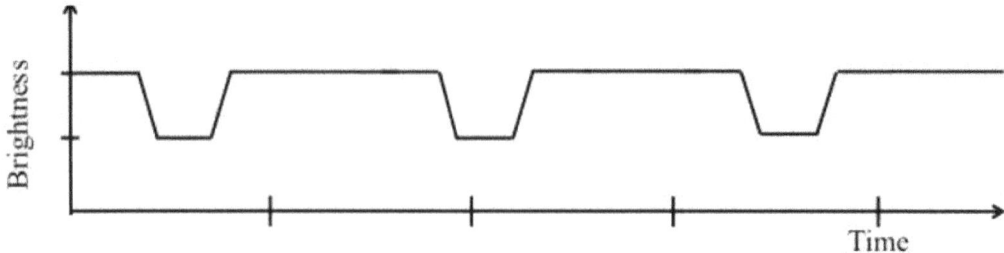

Figure 19.5: Light Curve Showing Multiple Transits of Single Exoplanet

When a star has more than one exoplanet, there will be more than one series of dips that add up. By looking at the plot of star brightness versus time, you can determine how many exoplanets there are and the relative sizes and orbital periods of the planets. Figure 19.7 shows an example of how this works. There are two exoplanets. One causes the pattern of dips that is shallower, the other causes the pattern of dips that is deeper. The deeper dips represent a planet that is bigger and has a shorter orbital period than the other planet. The pattern of dips will add up to produce the brightness curve shown at the bottom. Note that there are times when both planets are in front of the star at the same time; in this case, the dip produced is bigger than the dip the individual planets could have made.

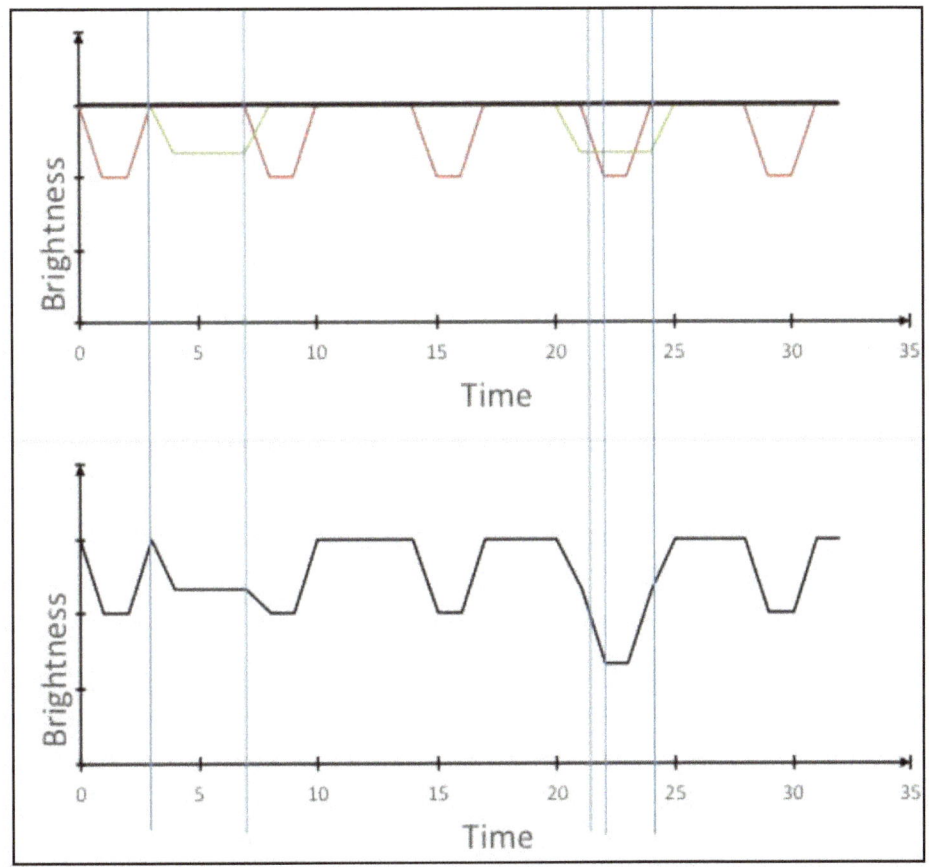

Figure 19.7: Light Curve for a Star with Two Exoplanets (CAE Astronomy Workshop Materials)

Questions You Should Be Able to Answer
1. What is the Doppler Effect? When is light from a star redshifted, and when is it blueshifted?
2. How is the radial velocity method used to detect extrasolar planets? What planetary properties can this method determine?
3. How is the transit method used to detect extrasolar planets? What planetary properties can this method determine?

Suggestions for Further Reading
1. http://astro.unl.edu/naap/esp/detection.html (summary of exoplanet detection methods)
2. http://astro.unl.edu/classaction/animations/extrasolarplanets/ca_extrasolarplanets_graph.html (radial velocity animation)

3. http://astro.unl.edu/classaction/animations/extrasolarplanets/transitsimulator.html (transit simulator)

Suggestions for Practice
4. http://astro.unl.edu/interactives/stellarvelocity/stellarVelocity1.html (radial velocity questions)
5. http://astro.unl.edu/interactives/doppler/DopplerShift2.html (doppler shift questions)
6. http://astro.unl.edu/classaction/questions/extrasolarplanets/ca_extrasolarplanets_radialvelocitycurve.html (radial velocity method questions)
7. http://astro.unl.edu/classaction/questions/extrasolarplanets/ca_extrasolarplanets_espdopplershift.html (more radial velocity method questions)
8. http://astro.unl.edu/classaction/questions/extrasolarplanets/ca_extrasolarplanets_ampperiod.html (more radial velocity method questions)

Chapter 20: Exoplanet Characteristics

In this chapter, you will learn about the characteristics of known exoplanets.

Let's look first at the sizes of exoplanets. Figure 20.1 shows that there are about 175 Earth-sized planets (less than 1.25 times the Earth's radius), about 300 Super Earth-sized planets (1.25 - 2 times the Earth's radius), about 600 Neptune-sized planets (2 - 6 times the Earth's radius), and about 700 Jupiter-sized planets (greater than 6 times the Earth's radius). This data makes it seem like there are more bigger exoplanets than smaller exoplanets. However, this is probably an observational bias...remember that bigger exoplanets are easier to find because they cause bigger transit dips and larger stellar radial velocities.

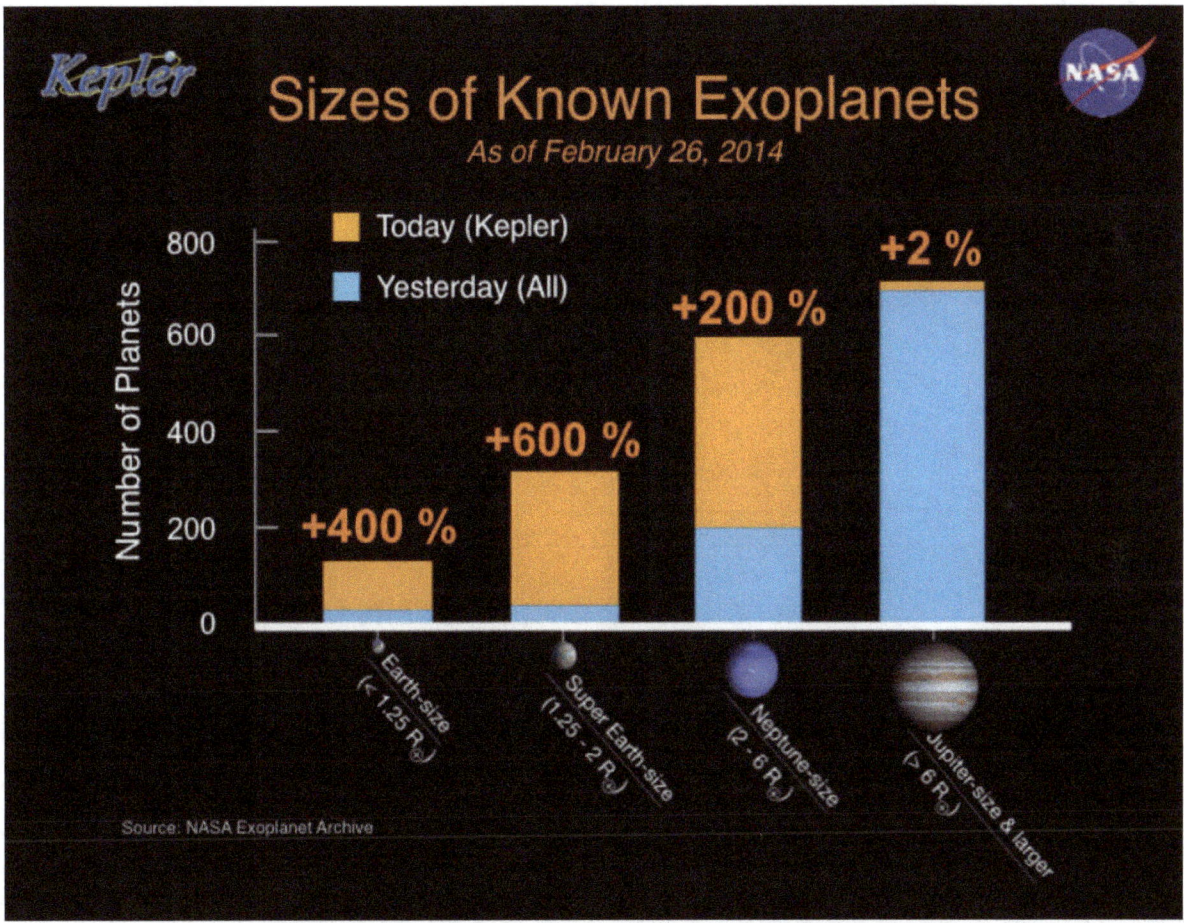

Figure 20.1: Exoplanet Sizes (http://www.nasa.gov, Kepler)

We would also like to look at how far these planets are from their stars. Figure 20.2 plots masses versus distances for all known exoplanets (as of 2014). Different colors represent different detection methods; for example, the planets in light gray were detected by Kepler. On the y-axis are the planets' masses in terms of Jupiter's mass. On the x-axis is the planets' distances from their stars in AU. Jupiter is 1 Jupiter mass and 5 AU, and so plots at the top star in the diagram. Thus planets similar to Jupiter in terms of mass and distance have been found. Earth is 0.003 Jupiter masses and 1 AU, and so plots at the bottom star in the diagram. You can see that though planets similar in size to the Earth have been found, we have not found planets that are similar in both size AND distance to the Earth. Again, this is probably an observational bias. Remember that planets that are close to their stars are easier to identify because they will transit their stars' more often. To find planets that are further away, a longer time period of data is needed, and we don't have enough data yet! (This research has made use of the Exoplanet Orbit Database and the Exoplanet Data Explorer at exoplanets.org.)

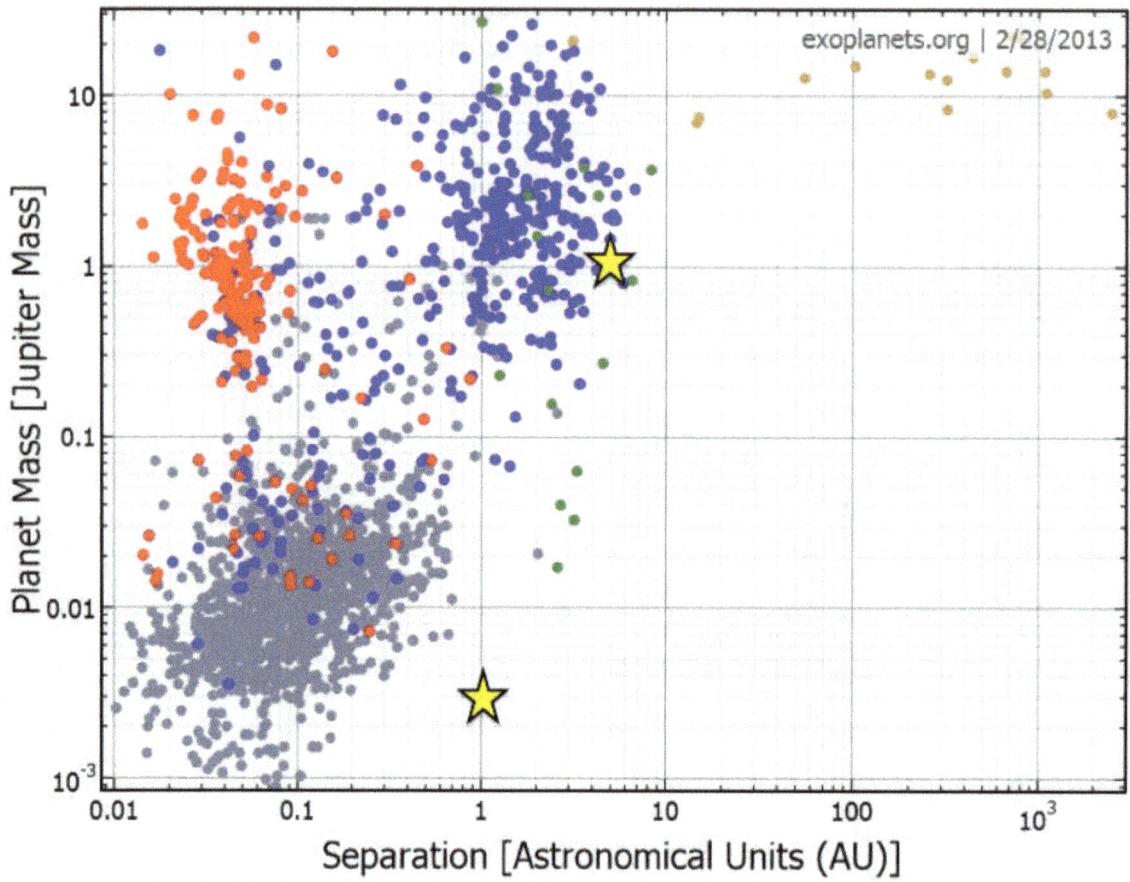

Figure 20.2: Mass Versus Distance of Exoplanets (http://www.exoplanets.org)[1]

We would like to know if any of these planets have the possibility of harboring life. Life as we know it on the Earth appears to need three essential ingredients: an energy source, organic molecules, and liquid water. Any planet orbiting a star automatically has an energy source (the star's energy). Likewise, organic molecules seem to be found pretty much everywhere. Therefore, the limiting factor is liquid water. Thus one way to start to answer the question of life on exoplanets is to look at whether liquid water could exist on these planets' surfaces. To this end, scientists define what is known as the habitable zone. The habitable zone around a star is the zone in which a planet could potentially have liquid water on its surface, provided that the atmosphere pressure is not too high or too low. The diagram in Figure 20.3 shows one estimates for the habitable zone around our Sun. Earth is at the inner edge of the habitable zone; depending on how optimistic you are, Mars may be at the outer edge (as it is in this diagram). The planets interior to the habitable zone are too hot for liquid water and the planets exterior to the habitable zone are too cold for liquid water.

---

[1] Eunkyu Han et al., "Exoplanet Orbit Database. II. Updates to Exoplanets.org," *Publications of the Astronomical Society of the Pacific* 126, no. 943 (2014): 827.

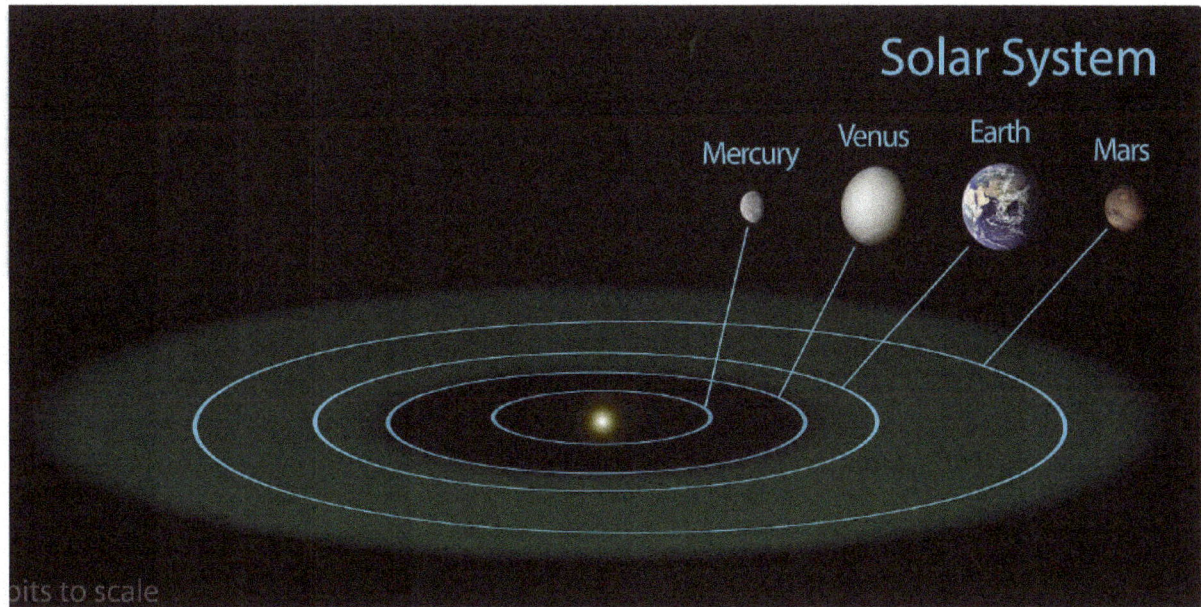

Figure 20.3: Estimate of the Habitable Zone in Our Solar System
(https://commons.wikimedia.org/wiki/File:Orbiting_in_the_Habitable_Zone_of_Two_Suns.jpg)

The distance of the habitable zone from its star depends on the size of the star (Figure 20.4). If a star is larger than our Sun, it will be hotter, moving the habitable zone further away. If a star is smaller than our Sun, it will be colder, moving the habitable zone closer.

Note that just because a planet is in the habitable zone does not mean it has to have liquid water; it just means that it could. In order to have liquid water, the size of the planet also needs to be within a certain range as well. If a planet is too small, then its interior will cool, its magnetic field will die, and its atmosphere will erode. This is what happened to Mars. If a planet is too large, then there will be too much H and He (it will be a Jovian planet), and the atmosphere will be too hot and the pressure will be too high for liquid water. It's a bit like Goldilocks and the Three Bears; the planet can't be too big, it can't be too small, it can't be too close to its star, and it can't be too far away from its star. It must be just right.

As of June 6, 2018, about thirty potentially habitable Earth-sized planets have been found. A milestone occurred in April of 2014, when the first Earth-sized planet in its star's habitable zone was found. This planet is named Kepler-186f.

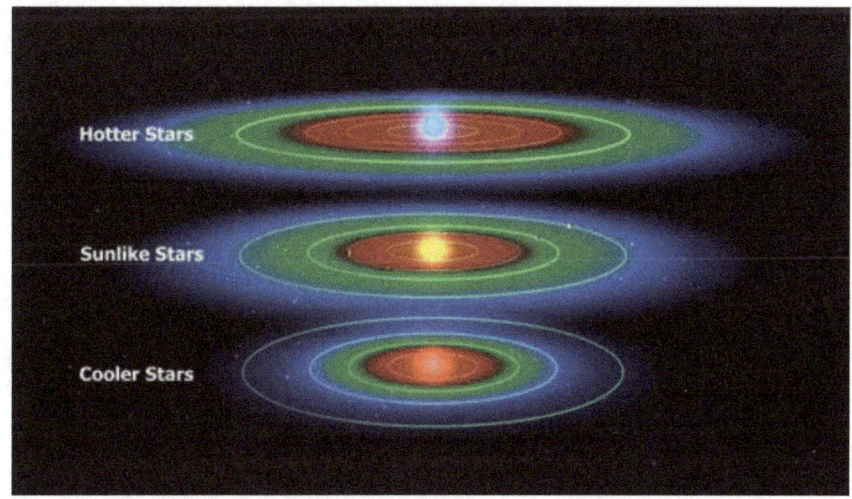

Figure 20.4: Habitable Zones Around Different Stars
(https://commons.wikimedia.org/wiki/File:HZ-by-size.jpeg)

Kepler-186f is about the same size as the Earth, as you can see in the to-scale comparison in Figure 20.5. Note that we do not actually have a picture of Kepler-186f – it is much too small to photograph. The image here (and virtually all images of exoplanets you will see in the news) is an artist's conception. The diagram in Figure 20.5 also compares the Earth's solar system with Kepler-186f's solar system. The star in Kepler-186f's solar system is smaller and colder than our star, and so Kepler-186f is located closer to its star than we are to the Earth. Kepler-186f is 0.36 AU from its star, orbits its star once every 130 Earth days, and is located about 500 light years away from the Earth. There are 4 other planets in Kepler-186f's solar system.

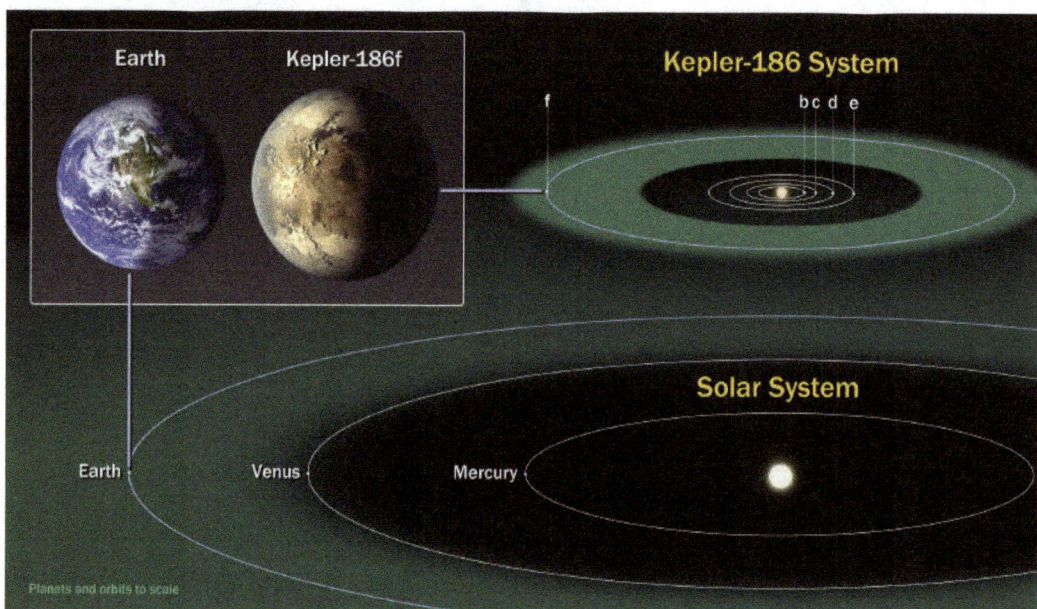

Figure 20.5: Kepler-186 System (http://www.nasa.gov)

We can use the statistics of known exoplanets to estimate the number of habitable planets. Statistics suggest that 1 in 6 stars have an Earth-sized planet, and that 5% of these are in the habitable zone. There are somewhere between 100 and 400 billion stars in our galaxy. This means that there should be somewhere between 17 billion and 67 billion Earth-sized planets in our galaxy and that there should be between 833 million to 3 billion habitable planets...just in our galaxy alone!!

Questions You Should Be Able to Answer
1. Are most exoplanets that have been discovered bigger or smaller than the Earth?
2. What is a habitable zone?
3. How does the size and distance of a habitable zone around a star depend on the temperature of the star?
4. Are habitable planets common in our galaxy?

Suggestions for Further Reading
1. http://planetquest.jpl.nasa.gov (NASA-JPL exoplanet website)

Reference
Eunkyu Han et al., "Exoplanet Orbit Database. II. Updates to Exoplanets.org," *Publications of the Astronomical Society of the Pacific* 126, no. 943 (2014): 827.

CPSIA information can be obtained
at www.ICGtesting.com
Printed in the USA
LVHW052151160120
643872LV00011B/574